Oil
A Beginner's Guide

ONEWORLD BEGINNER'S GUIDES combine an original, inventive, and engaging approach with expert analysis on subjects ranging from art and history to religion and politics, and everything in-between. Innovative and affordable, books in the series are perfect for anyone curious about the way the world works and the big ideas of our time.

Beginners
GUIDES

Oil
A Beginner's Guide

Vaclav Smil

ONEWORLD

A Oneworld Book

First published by Oneworld Publications, 2008
This new revised edition published 2017
Reprinted, 2020, 2022, 2023

ISBN 978-1-78607-286-3
eISBN 978-1-78607-287-0

Typeset by Siliconchips Services Ltd, UK
Printed and bound in Great Britain by
Clays Ltd, Elcograf S.p.A.

Oneworld Publications
10 Bloomsbury Street
London WC1B 3SR
England

Stay up to date with the latest books,
special offers, and exclusive content from
Oneworld with our newsletter

Sign up on our website
oneworld-publications.com

Contents

Contents

Figures

Unless otherwise stated the illustrations are original graphics prepared by Bounce Design in Winnipeg and, in the case of graphs, plotted from a variety of (historical and the latest) statistics.

Preface

Few short periods in the history of the oil industry have been as eventful as the one that followed the publication of this guide's first edition in 2008. The intervening developments have gone far beyond new rounds of oil price rises and collapses: there have also been notable technical advances in oil production, new economic realities affecting both major oil-producing and oil-consuming countries, and shifting perceptions of the global environmental prospects. This new edition preserves what has remained durable and fundamental, and updates the contents by including all notable post-2007 developments. The latest available data were always used, which means that most of it is carried through to 2016, while some coverage ends in early 2017. The new edition also takes a closer look at some key changes affecting the oil industry and oil's place in the modern world, above all at the recurrent peak oil claims, at the great American oil renaissance and the purported end of oil ('leave in the ground' sentiment) due to the necessity of accelerated decarbonization of the global energy supply required to prevent catastrophic global warming.

The overall approach has not changed: much like my republished book on energy, this brief book is not a guide for the beginner in the strictest sense. In both cases, a certain amount of basic scientific understanding (above all reasonable numeracy) is essential. The minimum entry-level for this book could be specified as an equivalent of the North American high school education; a year or two of university studies (no matter in what subject) would make for an easier read – but, as always, it is not formal qualifications but individual interest, inquisitiveness and willingness to learn that matter most. From that point of view readers who could profit from this book range from true beginners to people who know a great deal about a specific segment of the vast oil-centered enterprise but who would like to learn more about other aspects of this inherently interdisciplinary subject of scientific inquiry.

The book teems with numbers (I am sure too many for some tastes) but I make no apologies for this: real understanding of oil's origins, geology, exploration, extraction, transportation, processing, use and linkages to society and the environment can come only by appreciating the magnitudes of specific time spans, depths, volumes, durations, rates, cumulative totals, concentrations, prices, subsidies and costs that define and govern this vast global endeavor. As for the multitude of technical terms, I have tried to explain them (however briefly) whenever they are first used. All units and their abbreviations are listed in Appendix A, and Appendix B offers a dozen books for additional reading and a small selection of highly informative websites.

Revising the text was as enjoyable as writing the original, but, again, it was not particularly easy because for every interesting bit of information, for every number and for every conclusion that I have included I had to leave out several times that number of fascinating facts, explanations and useful asides pointing in unexpected directions. Squeezing the universe of oil into no more than 60,000 words of text remained an unending exercise in truncation and exclusion. And although this does not excuse all omissions and imperfections of the book, I ask both the experts (who might be incensed by the absence of matters they would have considered essential) and the true beginners (who would have wished for more extensive explanations) to keep in mind the fundamental restriction under which I had to labor. Finally, my thanks to Shadi Doostdar for giving me another opportunity to revisit the fascinating world of oil, and to Bounce Design for preparing the graphics.

World of oil

If history is seen as a sequence of progressively more remarkable energy conversions then oil, or more accurately a range of liquids produced from it, has earned an incomparable place in human evolution. Conversions of these liquids in internal combustion engines have expanded human horizons through new, and more affordable, means of personal and mass transportation. Anybody with a car in a country with decent highways can travel more than 1,000km in the course of one day (in Europe this could easily entail driving across four countries). Any city with a runway long enough to accommodate large jets can now be reached from any other city on the Earth in less than twenty hours of flying time, and for many people trips to Bali or Mauritius have become nearly as common as those to Birmingham or Munich. Liquid fuels have created new landscapes of concrete and asphalt highways, overpasses, parking lots, shopping megacenters and seemingly endless urban sprawl.

Private cars also allow for unprecedented quotidian personal mobility. They make it easy to buy imported foodstuffs in a store at the other end of a town or to drive, on the spur of the moment, to a restaurant, symphony concert or a football game. They make it possible to live far away from a place of work, to set one's own schedule during vacation drives, to spend free time far from home fishing or inside a garage installing monster engines and wheels or minutely reconstructing vintage car models.

Liquid fuels, through the combination of fast and massive container ships and eighteen-wheeler trucks, have brought us Chilean apricots and South African grapes in January and ginger or green beans from China or Kenya all year round. Liquid fuels have also helped to rationalize productive processes ranging from farming to retailing, changes that include such remarkable organizational feats as the just-in-time delivery of goods (where large assembly plants carry no extensive and expensive inventories and receive their parts by truck and train just when needed) and such profound macroeconomic changes as the globalization of manufacturing, where everything seems to be made (or assembled) many time zones away.

Modern life now begins and ends amidst the plethora of plastics whose synthesis began with feedstocks derived from oil – because hospitals teem with them. Surgical gloves, flexible tubing, catheters, IV containers, sterile packaging, trays, basins, bed pans and rails, thermal blankets and lab ware: naturally, you are not aware of these surroundings when a few hours or a few days old, but most of us will become all too painfully aware of them six, seven or eight decades later. And that recital was limited only to common hospital items made of polyvinylchloride; countless other items fashioned from a huge variety of plastics are in our cars, aeroplanes, trains, homes, offices and factories.

But if the new oil-derived world has been quasi-miraculous, enchanting and full of unprecedented opportunities, it has been also one of dubious deals, nasty power plays, protracted violence, economic inequalities and environmental destruction. Ever since its beginnings, the high stakes of the oil business have attracted shady business deals (from J. D. Rockefeller's Standard Oil to Mikhail Khodorkovsky's ill-starred Yukos) and begat some questionable alliances (be it the US and Saudi Arabia or China and Sudan). Oil ownership and the riches it provides have empowered dictators (from Muammar al-Gaddāfī to Saddām Hussein), emboldened autocrats (Vladimir Putin and the late Hugo Chavez being among the prominent examples), financed terrorists (including the murderous activities of *al-qā'ida* and *dā'ish*, the self-proclaimed Islamic State of Iraq and the Levant), encouraged massive corruption (be

it in Nigeria or Indonesia, Russia or Malaysia), promoted ostenta-
tiously excessive consumption (practiced by the legions of Saudi
princes as well as by new Russian oligarchs), engendered enormous income inequalities and done little for personal freedoms
and the status of women.

Many (perhaps too many) books about oil have looked
at these economic, social and political linkages. I will begin
by briefly examining oil in these contexts before going on to
explore the innumerable quotidian tasks of discovering, producing, transporting, refining and marketing the requisite volume
of oil, a mass that now amounts to well over 4 billion tonnes a
year. Once appreciated, these actions are no less fascinating than
the world of political oil intrigues, and only their cumulatively
immense ingenuity has made crude oil the single most important
source of primary energy in our world.

1
Oil's benefits and burdens

Dominant energies and the devices and machines used to convert them into heat and kinetic energy have left deep, and specific, imprints on society. The age of biomass energy relied on wood, charcoal and crop residues (known as biomass fuels) that were not always actually renewable as demand for heating and metal smelting often led to extensive deforestation and the overuse of crop residues. Small waterwheels and windmills powered by water and wind had a marginal role as human and animal muscles energized most tasks. The coal age introduced fuels that were more energy dense than wood, were available in highly concentrated deposits and in prodigious amounts from a relatively small number of mines, and could economically power steam engines. These were the first inexpensive mechanical prime movers that not only replaced many stationary tasks that had previously been performed by animal and human power, but also turned old dreams of rapid land and ocean travel into inexpensive realities.

The introduction and diffusion of refined oil products (gasoline, kerosene, diesel fuel, fuel oil) marked an even more important qualitative shift in modern energy consumption. New fuels were superior to coal in every respect: they had higher heat content (releasing more energy per unit mass when burned), were easier and safer to produce, cleaner and more convenient to burn and offered an incomparable flexibility of final uses.

Crude oil, or more accurately a variety of refined oil products derived from it, has changed the very tempo of modern life. By allowing the introduction of more efficient prime movers they increased the productivity of modern economies and they accelerated, as well as deepened, the process of economic globalization. Their extraction and sales have fundamentally changed the economic fortunes of many countries, and they have also improved some aspects of environmental quality and added immensely to private and public comfort. The nominal price paid for these benefits – the cost of finding crude oil, extracting it, refining it and bringing the products to the market – has been, so far, relatively affordable for all but the poorest of the world's economies.

The history of the oil business and of the price for crude oil paid by consumers are matters of rich documentary and statistical record and I will briefly recount major events, shifts and trends. But the prices that countries and companies pay for importing crude oil and the prices consumers pay when buying refined oil products (directly as automotive fuels and lubricants, indirectly as fuels for public and freight transport and for energy embedded in the production of virtually anything sold today) tell us little about the cost of finding and producing oil, and they are obviously very different from the real cost that modern societies have paid for oil in terms of (what economists so coyly call) the externalities of its extraction, transportation, processing and combustion, as well for ensuring the security of its supply.

That is why in the closing section of this chapter I will describe some of the broader costs of oil's benefits: the environmental consequences of energizing modern economies with liquid fuels ranging from marine oil pollution and photochemical smog to the combustion of refined products as major contributors of anthropogenic greenhouse gases; the economic, political and social impacts of both owning, and so frequently mismanaging, rich oil resources on the one hand and of being forced to buy them at what often amounts to extortionate prices on the other; and the political, military and strategic designs, calculations and decisions aimed at securing a steady flow of crude oil from the major producing regions and the wider repercussions of these activities.

What we have accomplished with oil

The beginnings of the oil era were not all that revolutionary: they started with a single product limited to just one major market as kerosene refined from crude oil became a major illuminant during the late 1860s and the 1870s. But it was not the only source of light, as city gas, made from coal, had been making great inroads in urban areas and soon afterwards both kerosene and gas were displaced by electricity. And neither the lightest nor the heaviest liquid fractions of crude oil were of much use in the early decades of the oil industry: gasoline was an inconvenient by-product of kerosene refining, too volatile and too flammable to be used for household lighting or heating, and there were no suitable small furnaces that could burn heavy oil for space heating. At least oil-derived lubricants offered cheaper and better alternatives to natural oils and waxes.

Only the invention of internal combustion engines (gasoline ones during the 1880s and the diesel engine during the 1890s) made oil's lighter fractions potentially valuable but they became indispensable only two decades later, and then only in North America, with the emergence of large-scale car ownership and the diffusion of trucking (elsewhere the conversion from railroad to highway transport and the rise of car ownership began only after World War II). Less than two decades after the first motor-ized vehicles came the use of gasoline-powered reciprocating engines in flight and, within a generation after this fundamental breakthrough, the emergence of commercial aviation after World War I. During the 1950s this new business was revolutionized by the introduction of gas turbines. These superior internal combustion engines made long-distance flight affordable.

Refined fuels powering massive diesel engines also changed both freight and passenger waterborne transport: all ships that were previously fueled by coal, from river barges to trans-oceanic liners, and from fishing vessels to large container ships (whose introduction made marine shipping a key tool of globalization) have benefited from the cleaner, cheaper, faster, more powerful and more reliable manner of propulsion. Small gasoline-powered outboard engines created a new leisure activity in motorized boating. Freight and passenger trains benefited from diesel engines, as did numerous heavy-duty trucks and construction and off-road vehicles.

Obviously, refined oil products have had their most far-reaching impact in transportation and I will note the key technical milestones of these advances and describe the current fuel requirements of these activities. The automobile was a European invention and its mechanical beginnings go back to 1876 when Nikolaus Otto (see figure 1) built the first four-stroke cycle engine running on coal gas. The first light, high-speed, gasoline-powered, single-cylinder vertical engine using Otto's four-stroke cycle was designed by Gottlieb Daimler and Wilhelm Maybach in 1885, and in the same year Karl Benz built the world's first motorized carriage powered by his slower horizontal gasoline engine. After a major redesign by Emile Levassor in 1891 the standard car configuration was virtually complete by the mid-1890s: the combination of four-stroke gasoline-fueled engine, electrical ignition and a carburetor launched the largest manufacturing industry in history whose expansion still continues.

An entirely different mode of fuel ignition was patented by Rudolf Diesel in 1892 (see figure 1). Fuel injected into the cylinder of diesel engines is ignited spontaneously by high temperatures generated by compressing the fuel twice as much as it is compressed in Otto's engines. Diesel engines work at a higher

Figure 1 Creators of the automobile age (clockwise): Nikolaus Otto, Karl Benz, Gottlieb Daimler and Rudolf Diesel

pressure and lower speed, and large stationary machines have best efficiencies just above 50% and automotive engines can approach 40%. Gasoline engines used to be 20–30% less efficient but their best new designs have almost closed the gap. Diesel fuel has other advantages: it contains about 11% more energy than gasoline in the same volume, it is slightly cheaper than gasoline and it is not dangerously flammable. Low flammability makes diesel engines particularly suitable in any setting where a fire could be an instant disaster (such as on board ships) as well as in the tropics where high temperatures will cause little evaporation from vehicle and ship tanks. And the combination of high engine efficiency, higher volumetric energy density and low fuel volatility means that diesel-powered vehicles can go farther without refueling than equally sized gasoline engines. Additional mechanical advantages include the diesel engine's high torque, its resistance to stalling when the speed drops, and its inherent ruggedness.

But early diesel engines were simply too heavy to be used in automobiles, and gasoline-fueled machines were not an instant success either: for more than a decade after Levassor's redesign (and also after Charles Duryea built the first American gasoline-fueled car in 1892) cars remained expensive, unreliable machines bought by small numbers of privileged experimenters. This changed only with Henry Ford's introduction of the affordable and reliable Model T in 1908 and with the expansion and per-fection of mass production techniques after World War I. Greater affordability combined with higher disposable incomes alongside technical advances in car design and better automotive fuels led to an inexorable rise in car use, first in the US, and then after 1950 in Europe and Japan, and now throughout much of conti-nental Asia.

The combination of America's affluence and perfected mass production gave the country a more than 90% share of the world's automotive fleet during the late 1930s, but the post-WWII eco-nomic recovery in Europe and Japan began to lower this share. In 1960, the US still had 60% of the world's passenger cars, but by 1983 Europe matched the US total and the continent is now the world's largest market for new vehicles while China became the fastest growing new car market during the 1990s. In 2015 worldwide passenger car registrations surpassed 900 million (see figure 2) and there were also about 350 million trucks, buses and cars in commercial fleets making a total of 1.25 billion road vehicles. Because the typical performance of their engines remains rather inefficient their claim on refined fuels remains high.

Any brief recital of the key economic, social and behavioral impacts of global car use must include, on the positive side of the ledger, unprecedented freedom of travel, expansion of individual horizons, flexibility and convenience, and the enormous contribution to the prosperity of modern economies where car

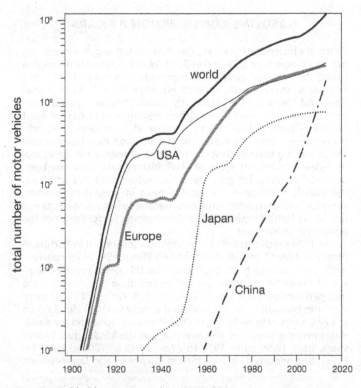

Figure 2 Worldwide car ownership, 1900–2015

building is commonly the single largest industry (in terms of added value) and where activities associated with the ownership and driving of cars create a large share of gross domestic product. The two lead items on the negative side are a large death and injury toll (worldwide, about 1.2 million deaths every year, and some twenty million injuries to drivers, passengers and pedestrians) and various environmental impacts. Traffic jams, now nearly chronic in most large urban areas, loss of land (often prime farmland) to highways and parking lots and the destruction of traditional urban patterns are other common negatives.

GASOLINE CONSUMPTION BY CARS

Thermal efficiency of the best gasoline-fueled engines in passenger cars is now over 30% and in 2014 Toyota developed an engine (using the Atkinson cycle) with maximum efficiency of 38%, but engines in everyday use achieve no more than 25%. Frictional losses cut the overall efficiency by about 20%, and partial load factors (inevitable during the urban driving that makes up most car travel time) reduce this by another 25%; accessory loss and (increasingly common) automatic transmission may nearly halve the remaining total so that the effective efficiency can be as low as 7–8%. Besides, for most of their history cars have not been designed to minimize gasoline consumption, and this has been particularly the case in the world's most important car market: America's preference for large cars, decades of low gasoline prices and heavy Detroit designs led to the declining performance of the post-WWII US car fleet.

In 1974 specific gasoline consumption (expressed in Europe in liters per 100km) actually increased by about 15% in comparison with the machines from the 1930s; the US uses a reverse measure of performance, miles per gallon (mpg), and hence this rate declined between the mid-1930s and 1974. Only OPEC's oil price increases brought a rapid turnaround as new federal rules (known as CAFE, Corporate Automotive Fuel Efficiency) specified gradually improving performance: the average was doubled in just twelve years, from 13.5mpg in 1974 to 27.5mpg (8.6 l/100km) by 1985. Expanding imports of more efficient European and Japanese cars

GASOLINE CONSUMPTION BY CARS (*cont.*)

further improved the overall performance. Unfortunately, the collapse of high oil prices in 1985 and then the economic vigor of the 1990s ended this desirable trend and CAFE remained stuck at 27.5mpg for the next 25 years, a huge loss of opportunity to make cars more efficient. Moreover, as pick-ups, vans and SUVs (sport utility vehicles: a monumental misnomer), all used primarily as passenger cars yet all exempt from CAFE standards, gained nearly half of the US car market, they dragged average fleet efficiency backwards.

The specific performance of these excessively large and powerful vehicles used to be mostly below 20mpg (above 11.8 l/100km), and some 2017 SUV models are even consuming more than 16 l/100km: Chevrolet's Suburban and Tahoe, and GMC's Yukon are in this monster category. Moreover, the stagnating efficiency was accompanied by a steady increase in average distance traveled per year: that rate barely moved between 1950 and 1975 (up by just 3% to 15,400km/vehicle) but by 2005 it had risen by more than a quarter to reach nearly 20,000km. As a result, average performance of all US light vehicles was still only 20mpg in the year 2000 and 21.5mpg ten years later. Then the average performance began, finally, to improve: by 2015 the mean had reached the record level of 24.8mpg and a number of bestselling cars could do better than 30mpg: the Honda Civic delivered 33mpg in combined city/highway cycle and 37mpg in highway driving, requiring just 6.35 l/100km.

Hybrid vehicles are, of course, much more efficient: the Ford Focus and Chevrolet Volt are just above 100mpg, requiring just 2.2 l/100km. And the combination of three trends – rising CAFE standards (by 2025 EPA stickers should be 43mpg for small vehicles and 37mpg for light trucks), further market inroads by popular hybrids, and a growing acceptance of electric vehicles – makes it very likely that even with slightly increasing car and truck fleets the US automotive gasoline demand may have already come close to its all-time peak (in 2016 it was just 0.1% above the previous record level set in 2007).

In 2016 motor and aviation gasoline accounted for a third of global refinery throughput. The US share of global gasoline consumption was about 41% of the total, or more than 1,200kg/capita: the country now consumes more gasoline than the combined total for the EU, Japan, China and India. The EU, with a population more than 50% larger than the US and with car ownership nearly

GASOLINE CONSUMPTION BY CARS (*cont.*)

as high as in the US, consumed only about 13% of the world's gasoline (about 160kg/capita). The key factors explaining this difference are the EU's higher number of diesel engines, smaller and more efficient gasoline-fueled vehicles, and much shorter average annual distances traveled by car (about half of the US mean). Japan consumed about 4% of the world's gasoline supply in 2015; China, with a population ten times larger than that of Japan, 10% (still only about 70kg/capita); and India claimed just 2%. These comparisons indicate the enormous potential demand for motor gasoline as car ownership increases in Asia's two most populous economies. They also make it clear that only major shifts in vehicle fleets (more efficiency, more hybrids and more electrics) will prevent this expansion from causing further serious deterioration of air quality.

The diesel engine has changed the world no less than its lighter but less efficient gasoline-powered counterpart. High weight/power ratio had delayed the use of diesels in passenger cars until after World War II but by the 1930s they were well on their way to dominating all applications where their higher mass made little difference, that is, in shipping, on railways, in freight road transport and in agriculture.

Just before World War II, one out of every four cargo ships was powered by diesel engines. Conversion to diesel accelerated after 1950 and today about nine out of ten freight ships are propelled by them, including the world's largest crude oil tankers and container vessels whose incessant traffic is the principal link between the producers and markets of the global manufacturing economy. The largest ships now have capacities closely approaching 200,000 deadweight tons (dwt, the weight of cargo plus ship's stores and bunkers and the fuel taken on board to power engines) and are able to carry more than 20,000 stacked containers at speeds exceeding 30km/h. Finnish Wärtsilä and Germany's Maschinenfabrik Augsburg-Nürnberg (MAN) are the leading designers of large marine diesels and Japan's Diesel United and South Korea's Hyundai are their leading producers.

Combustion of diesel oil has multiplied the energy efficiency of railway transport as the replacement of coal-fired steam locomotives by diesel engines boosted the typical conversion efficiency from less than 10% to at least 35%. Trunk rail lines everywhere are now either electrified or use diesel-powered traction.

Diesels began to replace gasoline-fueled vehicles in heavy road transport in 1924 when the first direct-injection diesel engine was made and when MAN and Benz and Daimler (two years before their merger) began to make diesel-powered trucks. By the late 1930s most of the new trucks and buses built in Europe were powered by diesel engines, and after World War II this dominance was extended to every continent. Diesels also power heavy-duty machines used in construction and surface mining, a variety of off-road vehicles (including trucks used in seismic exploration for oil), as well as those quintessential machines of modern land warfare, main battle tanks (although the US Abrams M1/A1 is powered by a gas turbine).

In 1926 Daimler Benz began to develop a diesel engine for passenger cars; their first model, a heavy saloon car introduced in 1936, became a favorite taxicab. Lighter, and also less polluting diesel engines were developed after 1950: consequently, the diesel engines in today's passenger cars are only slightly heavier than their gasoline-fueled counterparts and they should meet strict air quality standards. Although passenger diesels are still rare in North America (just 3% of new vehicles in the US), in Western Europe (with more expensive gasoline) diesel cars have accounted for slightly more than 50% of the new car market since 2006 (and in Ireland more than 70% since 2012).

A light gasoline-powered four-cylinder internal combustion engine built by the Wright brothers also powered the first flights of a heavier-than-air machine that took place at Kill Devil Hills, North Carolina, on December 17, 1903, after Wilbur and Orville solved the key challenges of balance and control and proper wing design by building a series of experimental gliders. Military planes powered by high-performance reciprocating engines saw plenty of action during the closing years of World War I and commercial flight began during the early 1920s, less than two decades after the Wrights' pioneering lift-off; by the late 1930s multi-motor hydroplanes were crossing the Pacific in stages. The performance of reciprocating aviation engines continued to improve until the late 1940s but their limits were clear: they had relatively high weight/power ratios; their action subjected the aeroplanes to constant vibration; they could not develop speeds in excess of 600km/h; and they could not sustain flight at high altitudes, above the often violent weather.

Prospects for long-distance commercial aviation changed fundamentally with the invention of jet engines and with their rapid adoption by airlines. Although the adjective is misleading, because the machines can burn both liquid and gaseous fuels, the proper technical name for jet engines burning kerosene is gas turbines. They are, much like the engines that power land vehicles, trains and ships, internal combustion engines but they differ from Otto and diesel engines in three fundamental ways. In jet engines the compression of air precedes the addition of fuel in a combustor, the combustion goes on continuously rather than intermittently, and the energy of the hot air flow is extracted by a turbine that is connected to the compressor by a shaft. Gas turbines first compress the air (up to 40 times above the atmospheric level) before forcing it through the combustion chamber where its temperature more than doubles. Part of the energy of the hot

GAS TURBINES IN FLIGHT

The construction of the first viable jet engine prototypes was a notable case of independent parallel invention as Frank Whittle in the UK and Hans Joachim Pabst von Ohain in Germany designed the first practical engines during the late 1930s. Von Ohain's version was first tested on August 27, 1939 in an experimental Heinkel-178, and Whittle's engine powered the experimental

GAS TURBINES IN FLIGHT (cont.)

Gloster on May 15, 1941. Improved versions of these engines entered WWII service too late (in July 1944) to make any difference to the outcome of the war.

Most of the great innovative military jet engine designs – driven by demands for ever higher speeds, altitudes and maneuverability – originated in the US and the USSR, but the British de Havilland Comet, powered by four de Havilland Ghost engines, became the first passenger jet to enter scheduled service, between London and Johannesburg, on May 5, 1952.

With a top speed of 640km/h the Comet was twice as fast as the best commercial propeller aeroplanes but it carried only thirty-six passengers and its engines had a very low thrust making it prone to loss of acceleration during take-off. But these drawbacks were not the reasons for the plane's catastrophic end. After three Comets disintegrated in the air between 1953 and 1954, all Comet flights were suspended and the fatal accidents were traced to the fatigue and subsequent rupture of the pressurized fuselage. When a completely redesigned Comet 4 began flying in October 1958, two other turbojets were in regular service, the Soviet Tupolev Tu-104 and the Boeing 707. The 707 was the first in a long line of the most successful commercial jet aircraft that includes the 737 (the bestselling jetliner in history) and the jumbo 747, the first wide-body jet (in scheduled service since January 1970). This enormous plane (maximum take-off weight of nearly 400t) was made possible by the development of turbofan engines.

By changing the gas compression and adding extra fans ahead of the compressor, two streams of exhaust gas are created – high-speed core exhaust which is enveloped by a volume of slower by-pass air; this reduces noise and produces a higher thrust. In the latest engine designs more than 90% of air compressed by an engine bypasses its combustion chamber, reducing both fuel consumption and engine noise. While turbojets reach their peak thrust at the very high speeds needed for fighter planes, turbofans do so at low speeds, a great advantage in making heavy planes airborne. The rapid post-1970 worldwide expansion of commercial flying would have been impossible without the low fuel consumption and very high reliability of turbofans (see figure 3). The engines are now so reliable that two-engine aircraft can be used even on the longest intercontinental routes on flights lasting seventeen hours.

gas rotates the turbine and the rest generates forward thrust by exiting through the exhaust nozzle.

Specific jet fuel (Jet A and Jet A-1) consumption (usually measured per passenger-kilometer) has been steadily, and impressively, decreasing, and the new Boeing 787 (Dreamliner) is nearly 70% more efficient than the company's pioneering 707 turbojet, whose commercial service began in 1958. Still, fuel consumption on long-distance flights is high: kerosene is 47% of the take-off weight of the Boeing 777–200LR, currently the passenger aeroplane with the longest-range; on a trans-oceanic flight nearly 45% (about 175t) of a Boeing 747 is kerosene and at cruising altitude (typically 10–12km above sea level) the aircraft's four engines consume about 3.2kg (roughly 4l) of the fuel every second. And absolute worldwide kerosene consumption has been rising steadily because air travel has seen the highest growth rate among all transportation modes. The annual total of passenger-km flown globally by scheduled airlines surpassed 40 billion in the early 1950s, and after doubling (on the average) in less than every six years, it reached nearly 3 trillion in the year 2000, and it surpassed 6 trillion in 2014 (see figure 3). In terms of passengers carried annually, the total rose from 320 million in 1970 to 3.4 billion by 2015. In 2015 the total mass of jet fuel consumed in commercial aviation was equal to only about 12% of the total of gasoline consumed by road vehicles, and jet fuel was less than 3% of worldwide refinery output, and in the US it was about 6% of total.

Nearly two-thirds of the world's refined products are now used in transportation (roughly 2.5Gt in 2005) and in the US that share is now more than 75%. Transportation's dependence on liquid fuels is even higher: in 2015 about 93% of all energy used by road vehicles, trains, ships and planes came from crude oil. Yet it can be argued that the most profound transformation effected by liquid fuels was the massive, and in affluent countries now pervasive, mechanization of agricultural tasks, a grand transformation of the most important economic activity that has been driven by a fundamental change of prime movers.

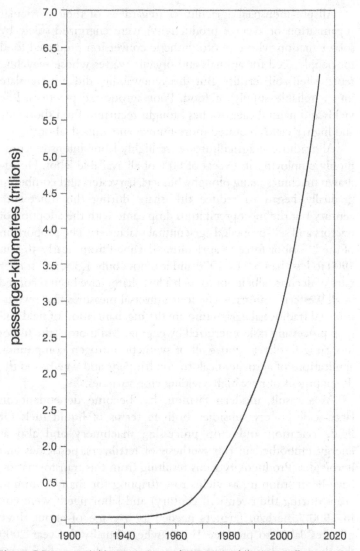

Figure 3 Exponential increase in total passenger-kilometers flown annually by scheduled airlines, 1920–2015

All pre-industrial agricultures (regardless of their particular organization or average productivity) were energized solely by solar radiation whose photosynthetic conversion produced food for people, feed for animals and organic wastes whose recycling replenished soil fertility. But this renewability did not translate into a reliable supply of food. Poor agronomic practices, low yields and natural catastrophes brought recurrent food shortages, and higher yields required more human and animal labor.

All traditional agricultures were highly labor-intensive, commonly employing in excess of 80% of all available labor. Horse-drawn machines (gang ploughs, binders, harvesters and combines) gradually began to reduce this share during the nineteenth century, but the most precipitous drop came with the adoption of tractors and self-propelled agricultural machinery. The proportion of the US labor force in agriculture declined from nearly 40% in 1900 to less than 5% by 1970 and it is now only 1.5%, and similar rates of decline (albeit not to such a low share) have been recorded in all Western countries. The four universal measures that revolutionized traditional agriculture are the mechanization of field and crop processing tasks energized by engines and motors; the use of inorganic fertilizers, above all of synthetic nitrogen compounds; applications of agrochemicals to combat pests and weeds; and the development of new high-yielding crop varieties.

As a result, modern farming has become dependent on large-scale energy subsidies, both in terms of liquid fuels for field, irrigation and crop processing machinery and also as energy embodied in the synthesis of fertilizers, pesticides and herbicides. Productivity gains resulting from this transformation have been stunning as yields rose (tripling for many common crops during the twentieth century) and labor needs were cut. In 1900 American farmers needed an average of about three minutes' labor to produce 1kg of wheat, but by the year 2000 the time was down to just two seconds and the best producers now do it in one second. The price of this progress is that, as

Howard Odum aptly put it, we are now eating potatoes partially made of oil.

Agricultural mechanization was made possible above all by the use of tractors, machines first introduced in significant numbers in the US just before World War I. The power capacity of gasoline-fueled tractors surpassed that of US draft horses before 1930, but in Europe the switch from animate to machine power took place only after World War II and it relied mostly on diesel engines that were introduced during the 1930s. Diesel engines also enabled the post-1950 shift to heavy four-wheel drive machines in the US and Canada (where the largest machines now rate about 400kW, or 550hp) as well as the designs of heavy caterpillar tractors. In contrast, Asia's agricultural mechanization has relied on small hand-guided two-wheel tractors (both gasoline and diesel powered) appropriate for small rice fields.

Diesel engines are also used in a variety of harvesting machinery, including self-propelled combines and cotton pickers. Stationary diesel engines of different sizes are also used to generate electricity, either in locations far from centralized electrical supply or in emergency situations, with the largest units as large as medium-sized steam turbines. Smaller engines are used to provide mechanical energy for refrigeration and crop processing. Given the magnitude of other final markets, agriculture (with forestry) makes a relatively small claim on refined products, amounting globally to less than 3% of the total.

Fuel oil was the first convenient substitute for solid fuels (coal and wood) whose combustion required repeated stoking and close supervision. After they bought small, automatically fed oil furnaces, millions of families, first in the US and Canada, later in Japan and Europe, could enjoy, for the first time, untended heat available at a flick of a switch or the setting of a thermostat. The small size of storage tanks means that delivery trucks must refill them four or even five times during a cold winter when high demand may force prices to spike. The worldwide switch to natural gas has reduced the number of families using fuel oil for space heating. In 2016 fewer than 5% of US households (about 6 million, about 90% of them in the Northeast) still relied on fuel oil compared to nearly a third in 1972, and this small share will soon be eliminated by abundant supply of natural gas produced by hydraulic fracturing of shales, a process to be explained in some detail later in this book. Worldwide, about 8% of all refined fuels were consumed by the residential and commercial sectors in 2015, overwhelmingly for heating; again, this share will continue to decline as natural gas takes over.

Besides supplying the most important liquid fuels of modern civilization the process of crude oil refining is also a source of key petrochemical feedstocks that are further processed into an enormous variety of synthetic materials. In 2015 about 11% of all hydrocarbon liquids (more than 400Mt) were used as petrochemical feedstocks, with naphtha accounting for about two-thirds of that total, followed by liquefied petroleum gases (LPG).

PETROCHEMICAL FEEDSTOCKS AND PLASTICS

There are two major kinds of these feedstocks, olefins (mainly ethylene and propylene) and aromatics (mainly benzene, toluene and xylene). Ethylene, produced by steam cracking of ethane or naphtha, is the most important petrochemical feedstock: the EU annually produces about 20Mt, the US about 40Mt. Propylene is the second most important feedstock and naphtha cracking also yields butadiene. Polymerization of basic feedstocks produces the now ubiquitous thermoplastics that account for more than 70% of all man-made polymers. Thermoplastics are made up of linear or branched molecules that are softened by heating but harden again when cooled.

Polyethylene is the most important thermoplastic, most commonly encountered as a thin but strong film made into garbage, grocery and bread bags, while its common hidden uses range from insulation of electrical cables to artificial hip joints. The material is also spun into fibers and blow-molded into rigid containers for

PETROCHEMICAL FEEDSTOCKS AND PLASTICS (*cont.*)

milk, detergents and motor oil, into gas tanks, pipes, toys and a multitude of industrial components.

PVC (polyvinyl chloride) is even more ubiquitous than polyethylene, found everywhere from buried pipes to credit cards, from floor tiles to surgical gloves.

Polypropylene is found in fabrics, upholstery and carpets. Propylene is also a starting material for such plastics as polycarbonates (in optical lenses, windows, rigid transparent covers and, when metallized, in CDs) and polyester resins.

Benzene is used in the synthesis of styrene (as polystyrene in packaging) and as a feedstock for a large number of other chemical reactions. Polyurethanes are a major end product of toluene and xylene is used in making polyester, solvents and films.

The second most voluminous non-fuel use of a refined petroleum product is asphalt. Asphalting of roads and sidewalks began sporadically in the US during the 1870s. New York City switched from brick, granite, and wood block paving to asphalt in 1896, and all of these early pavements were made with natural asphalt from Trinidad or Venezuela. Post-WWI car use increased demand for better pavings and the growth of the refining industry supplied hot mix asphalt derived from crude oil. The American experience, expanded after World War II with the building of the interstate highways, has since been repeated in all Western nations, and massive road building is now underway in China and in India, with concrete – a mixture of cement, water and aggregates – as the principal paving material.

But asphalt is easier to maintain: as long as the road foundations are sound the asphalt covering can be stripped and recycled. Indeed, asphalt, not aluminum cans or newspapers, is the most massively recycled material in affluent countries. In 2015 refineries worldwide produced nearly 90Mt of bitumen (a quarter of that in North America) and 85% of that output was used for paving to produce hot and warm asphalt mixes. In the US about 90% of the 70Mt of asphalt removed from worn surfaces in 2015 was reclaimed and reused in new hot and warm mix asphalt production. Asphalt is also used in roofing, industrial coatings, adhesives and in batteries.

Oil business and oil prices

In the past, multinational oil companies were seen as (and at times they actually were) prime practitioners of secretive, collusive, price-fixing deals. And even long after the pricing of oil slipped from their powers, they were still seen as manipulative and hardly trustworthy. When OPEC appeared to be in control of world oil prices, its national companies used to be viewed with even more distrust: whenever oil price spiked they were subject to unending references to how they were 'having us all over a barrel'. But things have changed and as recurrent rounds of major price declines have demonstrated the limits of OPEC's power, many commentators were quick to write the organization's (highly premature) obituaries. The realities have always been more complex. Since the early 1970s the oil business has been subject to a series of major fluctuations, with ups and downs beyond anybody's predictive powers, with troughs so deep that entire oilfields were shut down upon discovery and crests so high that surpluses and net profits broke all records.

During the 1970s and the early 1980s, major uncertainties were created by the lack of reliable information about remaining reserves, questions regarding the future revaluation of past discoveries and the near-panic concerns about any future moves by the seemingly omnipotent OPEC. The post-1985 retreat of crude oil prices eased such worries, and as relatively low prices fluctuated within a narrow range for the next fifteen years a welcome calm prevailed. In the early years of the twenty-first century, three new concerns converged to reignite the fears about the security of future oil supplies and crude oil prices. The declining rate of new oil discoveries, particularly as far as the new giant oilfields were concerned, misleadingly authoritative claims by some oil geologists (and by many so-called energy experts) about the imminent arrival of global peak oil extraction, and China's surging oil demand (as the country's economy continued to grow at near double-digit rates) led to a new round of rising oil prices. Their nominal peak in July 2008 was followed by a precipitous fall that was reversed again by new nominal highs in 2012 and 2013. They were nothing but a brief prelude to yet another swift retreat in 2014 and 2015 with low levels prevailing through 2016 and 2017. A somewhat different perspective arises when oil prices are expressed in constant monies (adjusted for inflation). This adjustment shows a century-long decline prior to 1970, and another period of general decline or stagnation between 1981 and 2003. After two spikes, in 2007–2008 and 2012–2013, the price in constant monies was no higher in 2015 than it was in 1985 (see figure 4). Historically speaking, the world price of crude oil continues to be a great bargain.

The impossibility of predicting sudden shifts in oil demand, the risk of violent conflicts and political upheavals, and a distorted understanding of oil resources combine to produce a market ruled by perception, fear, herd behavior and temporary panic – and hence commonly by overreaction. One year the headlines have the world drowning in oil, a few years later they have it facing the end of the oil era. All long-range forecasts of oil prices are thus irrelevant. OPEC's *Oil Outlook to 2025*, published in 2004, anticipated that oil prices would settle at between $20–25/b. But just a decade later OPEC's outlook changed, anticipating constant prices of $110/b until 2020 and then a slight decline to $100/b (in 2013 prices) by 2025. While OPEC was predicting the price to remain steady at $20/b it actually rose to $100/b, and when it thought it would stabilize at $110/b it actually fell to as low as $30/b, contrasts proving yet again the limits of OPEC's

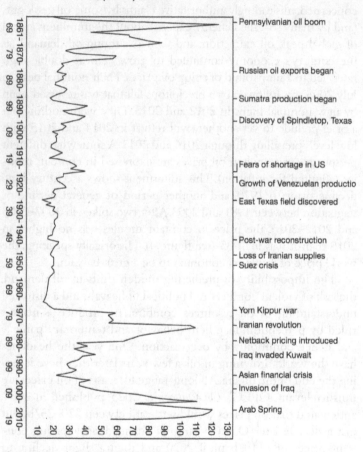

$(2016)/barrel

- Pennsylvanian oil boom
- Russian oil exports began
- Sumatra production began
- Discovery of Spindletop, Texas
- Fears of shortage in US
- Growth of Venezuelan productio
- East Texas field discovered
- Post-war reconstruction
- Loss of Iranian supplies
- Suez crisis
- Yom Kippur war
- Iranian revolution
- Netback pricing introduced
- Iraq invaded Kuwait
- Asian financial crisis
- Invasion of Iraq
- Arab Spring

Figure 4 Crude oil prices, 1861–2016

influence and the utter futility of even short-term forecasts, a fact that, alas, leaves no impression on small armies of economists engaged in constant price forecasting!

Basic facts about the global oil business involve some very large aggregates. In 2016 (when the average oil price was $43/b) global sales of crude oil were worth just over $1.5 trillion. This was equal to 2% of the world's economic product of about $75 trillion, a bit less than the GDP of Canada and almost equal to Russia's GDP. These comparisons confirm that crude oil is not an overpriced energy source. Matters change when we move (using oil industry parlance) downstream as oil companies add a great deal of value (profit) by transporting and refining crude oil and by marketing the final products, and governments also step in to collect taxes. Even then refined products are still quite afford-able in all high-income countries, but in the Netherlands in 2017 a barrel of gasoline was five times more expensive than a barrel of crude oil, with multiples ranging from less than 4.5 in the UK to about 3.5 in Japan, and to less than 2 in the US.

Despite world oil prices remaining relatively low, in 2016 five of the world's ten largest publicly listed companies (by annual revenue) were in the oil business: China National Petroleum Corporation (number 3, $299 billion), Sinopec (number 4, $294 billion), Royal Dutch Shell (number 5, $272 billion), Exxon (number 6, $231 billion) and BP (number 10, $226 billion). Combined revenues of these five companies reached about $1.3 trillion in 2016, surpassing the nominal annual GDP of Russia in that year. But in 2014, when the average world oil price was close to $100/barrel, the combined revenue of these five companies was nearly $2.1 trillion. By this point, oil had earned a reputation for such volatility. In 1998 *The Economist* headlined the oil industry as 'The decade's worst stocks.' This was followed by record profits at the turn of the twenty-first century that ended abruptly in 2008 with the greatest economic crisis since World War II. But price fluctuations aside, it must be remembered that, given the high taxes imposed on refined products by Western governments, oil producers have actually made substantially less money than state treasuries.

The US has been an exception: with federal and state taxes on gasoline amounting to 21% in 2016, the unit price of that fuel is mostly the cost of crude oil (about 43% in 2015) and the industry margin (about 36%). But in 2016, Japan's taxes amounted to 40% of the price of gasoline, while in Germany it was 55% and in the UK, 67%. As a result, the taxes collected on liquid fuels by the world's seven largest economies (G7) have surpassed the annual oil revenue of the 13 OPEC nations combined. In addition, since the 1960s, the global oil business has been dominated by state-owned companies whose role and revenue will only increase, as they control most liquid oil reserves and a large share of non-traditional reserves in oil (tar) sands.

NATIONAL OIL COMPANIES

These companies now control most of the world's oil reserves, and hence most of today's, and future, production. In 2015 the four largest ones – Saudi Aramco, National Iranian Oil Company (NIOC), Iraq National Oil Company (INOC) and Kuwait Petroleum Company (KPC) – held almost 40% of the world's total conventional oil reserves, virtually all of them in conventional (liquid oil) deposits. The next four – Petróleos de Venezuela, Abu Dhabi National Oil Company (ADNOC), Libyan National Oil Company and Nigerian National Petroleum Company (NNPC) – controlled an additional 28%, mainly thanks to Venezuela's huge deposits of Orinoco tar sands that make the country the world's first in total (conventional and non conventional) crude oil reserves, ahead of Saudi Arabia. Canada, thanks to Alberta's oil sands, now ranks third. When including publicly traded companies in the list, Russia's Rosneft, the country's largest firm, would come twelfth with about 1.8% of global oil reserves, followed by Exxon with nearly 1.5% of the total.

National oil companies have differed greatly in terms of their competence, performance and foresight. Norway's Statoil may be, in many ways, a model state oil company with transparent

NATIONAL OIL COMPANIES (*cont.*)

operations and extensive investment in oil exploration and production, but most of the state-run oil companies in modernizing
countries have been poorly managed and perform well below
their potential. The largest one, Saudi Aramco, headquartered in
Dhahrān on the Gulf, has been the world's largest oil producer
since 1978 when it completed its compensated nationalization of
Aramco's assets. The company has been run fairly smoothly but in
a secretive manner. That will certainly change in the future: in 2016
the company decided to proceed with the initial public offering of
up to 5% of its assets (at a predicted price of about $100 billion)
but by mid-2017 its actual timing and success remained uncertain.
Saudi Aramco's 2015 *Facts and Figures* and *Annual Review* publications contain data on reserves, production and refining capacity,
but only one table on sales (of refined products for the domestic
market) and not a single figure regarding the company's revenues
(in 2014, before average world oil prices fell by more than half,
those revenues were slightly more than a billion dollars a day) or
its operating costs or profits. The company has also released very
little information about the recent status of al-Ghawār, the world's
largest oilfield. This led to some speculation about the field's long-
term prospects but the continued high Saudi output clearly indicates that the field is not nearing its exhaustion. In 1973, the year
of the first round of large crude oil price increases, the company
produced 384Mt of crude oil and by 1980 the output had risen to
509.8Mt; the subsequent collapse of oil demand reduced extraction to as little as 172Mt in 1985 but a new record was set only
twenty years later with 521.3Mt in 2005, and in 2016 Saudi Arabia
produced about 586Mt, just 1.3% more than the US.

The National Iranian Oil Corporation (NIOC) controls about
9% of global oil reserves but for decades it was managing a modern industry in a country that did not allow foreign companies to
own equity in Iranian companies or to operate production concessions. The NIOC could thus only award contracts that guaranteed a
share of eventual production from a field developed with foreign
investment. Companies from Malaysia, France, Italy, Spain and
China have participated in these arrangements. This has changed
with the lifting of economic sanctions in 2016, after which the
NIOC invited Western companies to bid on oil and gas projects.
Output still has far to go before reaching the old record levels:
under the Shah during the mid-1970s Iran's oil flow peaked at
303Mt in 1974, but in 2016, at 216Mt, it was still nearly 30% lower
and barely higher than a decade ago.

A free market has not been one of the hallmarks of the 150 years of oil's commercial history. The oil business has seen repeated efforts to fix product prices by controlling either the level of crude oil extraction or by dominating its transportation and processing, or by monopolizing all of these aspects. The first infamous, and successful, attempt to do so was the establishment of Standard Oil in Cleveland in 1870. The Rockefeller brothers (John D. and William) and their partners used secretive acquisitions and deals with railroad companies to gain the control of oil markets first in Cleveland, then in the Northeast, and eventually throughout the US. By 1904 what was now known as the Standard Oil Trust controlled just over 90% of the country's crude oil production and 85% of all sales.

The trust was sued by the US government pursuant to the Sherman Antitrust Act of 1890 but it wasn't until 1911 that the order to dissolve it was upheld by the Supreme Court. The dissolution produced more than thirty separate companies that continued to use the Standard name, and the names of the largest of these are still prominent – after repeated mergers, reorganizations, acquisitions and name changes – among the world's largest publicly traded oil companies.

Standard of New Jersey became Esso. Standard of New York (Socony) merged with Vacuum Oil Company and in 1966 it was renamed Mobil. Esso was renamed Exxon in 1972 and in 1999 it combined with Mobil to form one of several double-name oil companies, ExxonMobil. Standard Oil of California (Socal) became Chevron in 1984 and in 2001 it merged with Texaco to form ChevronTexaco (with the Texaco brand remaining only outside North America). Standard of Ohio (Sohio) was bought by BP (between 1984 and 1987) and Standard of Indiana Amoco (rebranded as Amoco in 1973) merged with BP in 1998, but the double name BPAmoco lasted only until 2000 when BP also bought Atlantic Richfield (ARCO).

This longevity of major oil companies extends beyond the Standard pedigree and beyond the US business. In the US, Gulf

Oil was established in 1890, Texaco was set up in 1901, Royal Dutch Shell was chartered in 1907, and the Anglo-Persian Oil Company (using the name British Petroleum since 1917) was set up in 1909. In 1928 the chairmen of Standard Oil's three largest successors, Esso, Socony and Socal, and their counterparts from Royal Dutch Shell and Anglo-Persian met in Scotland's Achnacarry Castle, essentially to divide the global oil market and to stabilize the price of crude oil. After this informal oligopoly was joined by Gulf and Texaco it became widely known as the Seven Sisters (*le sette sorelle*, the name used first by an Italian oil-man, Enrico Mattei). Their domination of the entire chain of the global oil business, from exploration to gasoline marketing, made it possible to set prices for the newly discovered oil that they began to produce after World War II in the hydrocarbon-rich countries of Asia, Africa and Latin America.

The dominance of major multinational oil companies began to weaken during the 1960s with the rise of OPEC, and their global importance was rapidly reduced to a small fraction of their former strength by a wave of nationalizations during the 1970s. In 1960 the Seven Sisters produced more than 60% of the world's oil, but by 1980 that share had fallen to about 28%, and in 2016 it had declined further to only about 13%. OPEC was not the first organization set up explicitly to manage oil prices: it was modelled on the Texas Railroad Commission, a state agency that began to regulate railroads in 1891 and added responsibilities for the regulation of the oil and gas industry in 1919. After the discovery of the East Texas field brought a precipitous fall in oil prices, the commission was given the right, in 1931, to control the state's oil production through prorated quotas in the form of a monthly production allowance that set the permissible percentage of maximum output. With Texas being the country's largest oil producer and the US dominating global oil output (with Texas producing more than half of the world's crude) this right amounted to a very effective cartel of worldwide importance run by an obscure state agency. In 1950 the US still produced about 53% of the world's crude oil, a higher share than OPEC had at any time after 1973 (in 2015 its share was about 42%), but matters began to change radically during the 1950s.

Between 1950 and 1970 oil from new Middle Eastern discoveries began to reach the global market and helped to drive a worldwide economic expansion that proceeded at an unprecedented annual rate of nearly 5%. During that period US oil demand, relatively high to begin with, nearly tripled, and as post-WWII Western Europe and Japan began to convert from coal to oil-based economies their oil demand rose even faster so that by 1970 the affluent countries were consuming four times as much oil as they did in 1950. The beginnings of industrialization in many low-income Asian and Latin American countries further added to the rise in global oil demand. But new discoveries easily supported this and in 1960 major oil producing companies (led by the Seven Sisters) reduced their posted crude oil prices, the fictitious valuations that were used for calculating the taxes and royalties owed to the oil-rich nations whose resource these companies were selling worldwide. In response to this move five oil-producing states set up the Organization of Petroleum Exporting Countries (OPEC) in Baghdad in 1960.

As of 2017 OPEC had 14 members. Saudi Arabia, Iraq, Kuwait, Iran and Venezuela were the founding nations. Qatar joined in 1961, Libya and Indonesia in 1962, Abu Dhabi in 1967, Algeria in 1969, Nigeria in 1971, Ecuador in 1973 and Gabon in 1975 (the last two countries left the group in, respectively, 1993 and 1996). Angola came on board on January 1, 2007, the same year Ecuador rejoined and Indonesia left after its oil output declined. Seven years later, in December 2015, Indonesia returned only to be suspended a year later because of disagreements about production quota, Gabon rejoined, and Equatorial Guinea joined in 2017. The first step for the new organization was to protect its revenues: all the early OPEC members agreed not to tolerate any further reductions of posted prices, and income tax became an excise tax. By the late 1960s continuing high demand for oil had begun to create a seller's market. In response to this, in September 1971, Libya increased both its posted oil price as well as the tax rate paid by foreign oil companies. In February 1971, twenty-two leading oil companies accepted OPEC's demand (justified by a weaker dollar) for a new 55% tax rate, an immediate increase in posted price and future price increases.

Even before this took place rising oil demand led the Texas Railroad Commission to lift its limits on production in March 1971, ending an era of price-controlling power thanks to the newly assertive OPEC. Concurrently, the prospect of higher oil prices began a wave of nationalizations that continued for most of the 1970s: Algeria nationalized 51% of French oil concessions in February 1971; Libya began its nationalizing with BP holdings in December 1971; Iraq took over all foreign concessions in June 1972; OPEC approved a plan for 25% government ownership of all foreign oil assets in Kuwait, Qatar, Abu Dhabi and Saudi Arabia in October 1972; and in January 1973 Iran announced that it would not renew its agreements with foreign companies when they expired in 1979. Another important change that opened the way to a new regime of global oil pricing took place in April 1973 when the US government ended the limits on the import of crude oil east of the Rocky Mountains set by President Eisenhower in 1959. This decision saw a rapid rise in US oil imports.

On October 1, 1973, OPEC, looking for higher profits, raised its posted price of $2.59/b by 16% to $3.01. On October 16, 1973, following the Israeli victory over Egypt in Sinai, the six Arab Gulf states raised posted prices by an additional 17% to $3.65/b, and three days later OPEC's Arab members embargoed all oil exports to the US until Israel pulled out of occupied Arab territories (the embargo was soon extended to the Netherlands because Rotterdam had Europe's largest oil terminal and refineries). On January 1, 1974, the six Gulf states raised their posted price to $11.65/b, a 4.5-fold rise in one year. The embargo on imports to the US was abandoned in March 1974 (it could not succeed as multinational oil companies simply rerouted their tankers). OPEC's first round of steep price increases was followed by a few years of minor changes. In 1978 the price reached $12.93/b, and the Saudi government acquired complete control of Aramco, creating the world's largest national oil company (in 2015 it produced more than four times as much crude oil as Exxon, the world's largest private operator). The effects of more than quadrupling the world oil price between 1973 and 1974 (and quintupling it, in nominal terms, between 1973 and 1978) were rapid and far-reaching. In North America and Europe the sudden price rise and the embargo resulted initially in a (false) perception of a physical shortage of oil and led to long car queues at filling stations, fuel rationing schemes and widespread fears of being at the mercy of greedy OPEC countries in general and unpredictable oil-rich Arab regimes in particular.

These fears soon subsided (there were no fuel shortages) but the serious economic consequences of the large price hike became clear as consumers and national economies, habituated to decades of low (and in real terms falling) oil prices had no choice but to pay five times as much for fuel. The full impact on the US economy (in 1974 the US imported about 22% of its crude oil demand) was delayed because of the crude oil price controls that were imposed in August 1973 during President Nixon's second term. Consequently, the average inflation-adjusted price of US gasoline in 1978 was still no higher than it was a decade earlier, and the price of refined products rose to levels unseen since World War II only after the controls were abolished on January 28, 1981, when President Reagan came into office.

Japan (with all but 0.1% of its oil imported) and most European countries (importing in excess of 90% of their oil needs) were much more vulnerable than the US but they had one important advantage: their overall energy use was already much more efficient than in North America and the price shock only intensified these efficiency efforts and led to a higher reliance on other fuels and on nuclear electricity. Most remarkably, Japan's GDP, after falling by 0.5% in 1974, was up by 4% in 1975 even as the country's overall energy use fell by nearly 5%. OPEC's windfall was large: the total revenues of its member states tripled between 1973 and 1978, but high inflation generated by quintupled oil prices meant that between 1974 (after the initial hike) and 1978 world oil prices actually fell in real terms.

But a second round of oil price rises was about to begin. Demonstrations against Shah Mohammad Reza Pahlavi began in Tehran in January 1978. By the end of the summer Iran was under military rule and by December its oil production had fallen sharply. On January 16, 1979, when the Shah fled into exile, OPEC's oil price averaged $13.62/b; twelve months later, with Ayatollah Khomeini back in Iran, and with the US embassy occupied by student radicals, the price nearly doubled to $25.56/b. A year later (after Iraq invaded Iran in September) it was $32.95/b. The peak was reached in March 1981 with the average at $34.89/b, the best-quality crude oils selling on the spot market for around $50/b, and experts widely predicting prices of $100/b in just a few years.

Economies that had begun to recover from the first price hike were hit again, and more seriously. In 1982 the US GDP fell by 2%, but the record high oil prices caused the greatest setbacks in Asia, Africa and Latin America: their industries, transportation and also urban cooking (using kerosene stoves) depended on oil imports and the high (dollar denominated) prices were reducing their future export earnings. But with the second round of price rises OPEC clearly overplayed its hand. This time oil prices rose high enough to do three things that greatly weakened OPEC's dominance of the market. The resulting economic slowdown depressed the global fuel demand: by 1983 it was 10% below the 1978 peak (and in the US the cut was 21%); it reinforced the drive for higher energy efficiency; and it led to vigorous oil and gas exploration and development in non-OPEC countries. The results of this combination were impressive. In 1978 non-OPEC oil producers (excluding the USSR) extracted 35% of the world's oil but in 1983 their share rose to 45% while OPEC's share fell to just 31%.

At first OPEC tried to keep the prices high, lowering the market rate to only $33.63 per barrel during 1982, but the oil glut persisted and by early 1983 it had to cut it to $28.74/b. The end came in August 1985 when the Saudis decided to stop acting as the swing producer (repeatedly cutting their oil output to prop up the falling prices) and linked their oil price to the spot market values, and at the beginning of 1986 doubling their extraction in order to regain their lost market share. Oil prices fell to $20/b by January 1986, and in early April they dipped below $10/b before they temporarily stabilized at around $15/b. They remained low for the remainder of the 1980s and even the Iraqi invasion of Kuwait (on August 2, 1990) and the First Gulf War (January 16 – February 28, 1991) produced only short-lived spikes followed by a decade of prices that stayed mostly between $15–20/b.

The average export price reached $23/b in January 1997 but the drop in demand caused by a short but severe Asian financial crisis depressed the price to just $9.41/b by December 1998. Once again, low oil prices were being taken for granted and, once again, energy demand began rising even in those rich nations that were already by far the largest users, and importers, of oil. During the 1990s, energy consumption rose by almost 15% in the USA, 17% in France, 19% in Australia and, despite a stagnating economy, by 24% in Japan. OPEC's share of global oil output rose again to above 40% and oil prices rose to more than $25/b by the end of 1999 and briefly surpassed $30/b in September 2000. Prices fell again with the onset of the world-wide economic recession in the wake of the terrorist attack on the USA on 11 September, 2001, but they went on to double to more than $50/b by the end of 2005 and remained at that level in 2006. What followed has no precedent in the history of oil prices. Twice during the next ten years, a rapid ascent was followed by an equally rapid price drop.

THE DECADE OF UPS AND DOWNS

Crude oil prices (the following values refer to the US benchmark, West Texas Intermediate crude, FOB at the main trading hub, Cushing in Oklahoma) continued their steady rise and ended 2007 at $95/b, and then their growth accelerated to reach, during the first week of July 2008, a new all-time high of $145.31/b. On September 15, 2008, Lehman Brothers filed for bankruptcy and the worst economic crisis of the post-WWII era began to unfold at a shocking speed. By the end of 2008 crude oil was trading at just $33/b, a 79% slump in six months. As the US government began the economic rescue (by printing unprecedented quantities of dollars) prices began to recover: they more than doubled during 2009 and ended 2011 at $100/b. They went temporarily even slightly higher in 2012 and 2013 when they ended at $99/b. Then, in 2014, there was a close replica of 2008: the late June peak of $107/b was followed by a retreat to $53/b at the year's end. A year later the price drifted down to $37/b and in the second week of February 2016 it went as low as $26.19/b. By the end of the year it had doubled to $54/b but by mid-2017 it was once again below $50/b.

Several developments combined to produce this decade-long roller coaster. The most important factor behind the first rise was the increase in demand, both by traditionally large Western oil importers and by the rapidly expanding Chinese and Indian economies. But the price rise was also driven by the unsettled situation in occupied Iraq and by concerns about terrorist attacks in general and strikes on Saudi oilfields or oil terminals in particular. Moreover, low stock market returns attracted speculative investments in oil futures, and all of this while the media were disseminating misleading stories about an imminent peak of global oil production that was to be followed by a scramble for diminishing resources heralding the painful end of modern civilization (see chapter 5). In July 2008, as the price passed $140/b, there was no shortage of energy experts forecasting an imminent rise to $250/b – but they would have been wrong even without any economic downturn

THE DECADE OF UPS AND DOWNS (*cont.*)

because such a high price would have done what a similarly high price (when adjusted for inflation and oil intensity of the economy) began to do in 1981: destroy the demand.

At the same time, it must be realized that not only small changes in the global supply or demand, but their mere anticipation, can bring disproportionately large price moves and that there is no simple correlation between the two trends. In 1980 crude oil prices rose by 51% (driven by the takeover of Iran by fundamentalist mullahs) even though consumption fell by 4%. In 1986 consumption rose by 3% as prices fell by 46%; similarly, in 2009, consumption fell by nearly 2% but prices declined by 38%; and in 2015, consumption rose by nearly 2% but the price declined by 30%, with the rising US output, propelled by shale oil, creating fears of enormous supply gluts.

Nobody is in control of oil prices – or else that entity would have a peculiar taste for wild swings and a near-permanent lack of stability and predictability (see figure 5). The key lessons from this high price volatility and the global economic consequences of these unpredictable, excessive fluctuations have been widely ignored or misinterpreted. In the first place, the price gyrations in general, and the recurrent periods of high post-1973 prices in particular, have never reflected any imminent or rapidly approaching physical shortage of oil, as the resource remains abundant. For more than three decades a key reason for price over-reaction to small supply or demand moves was the minimal safety cushion created by OPEC: its production in 2003 was just 1% higher than in 1973, the year of the first round of oil price increases. By 2015 OPEC, in an attempt to break the rising power of US shale oil producers, by forcing prices lower and hoping that many of them would go bankrupt, was producing 20% more than the 2003 level, but this new strategy of flooding the market did not bring the desired effect and by November 2016 OPEC was, once again, cutting its crude oil production.

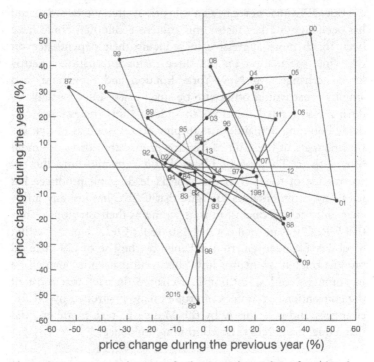

Figure 5 Year-to-year changes of oil prices, shown here for thirty-four years between 1981 and 2015, reveal virtually random shifts that are impossible to forecast.

The greatest challenge for OPEC has been always to keep the price below the level that would lead to a substantial drop in demand for oil, to increased hydrocarbon exploration in non-OPEC countries and to government-subsidized investment toward alternative energy sources. In its public pronouncements OPEC has repeatedly professed its aversion to such demand-destroying prices and its commitment to a stable oil market and security of supply, but its actions have often had the very opposite effect.

But OPEC has never been a sole price setter: Western demand has been always a key factor and affluent economies could have been much more aggressive in reducing their dependence on oil, while speculation on the three major international petroleum exchanges (in New York, London and Singapore) can amplify what would otherwise be small price shifts, particularly during exaggerated reactions to sudden shifts (recessions, suddenly booming demand), catastrophic events (such as Hurricane Katrina that cut the Gulf of Mexico production) or to the mere fear of them. The rise of US shale oil extraction and the re-emergence of the US as the world's leading oil producer has led many commentators to argue that OPEC has lost any influence, but such a conclusion is as wrong as their previous claims that OPEC was in total control. No doubt OPEC's price-setting capability has been greatly weakened by the rise of US shale oil production, but no matter how successful that endeavor will be in coming decades, nothing can change the fact that most of the conventional resources of crude oil are controlled by OPEC countries, and particularly by the Middle Eastern members, and that alone guarantees their continuing influence.

Oil links and the real cost of oil

The world's single most important source of fossil energy and its truly worldwide extraction, transportation, processing and combustion affect every realm of modern life. The performance of all but the poorest economies, matters of both domestic and international politics in both oil exporting and oil importing countries, quality of life, a great deal of strategic thinking on the part of major powers, particular military actions during times of war and the state of the Earth's environment – all of these are demonstrably linked to oil, but virtually all of these linkages are complex and their consequences are often counterintuitive.

Primary energies of fossil fuels have been the necessary engines of modern economies but their abundance alone is not enough to bring admirable economic achievements and to guarantee an improving standard of living.

Just before its demise the USSR was the world's largest producer of both crude oil and natural gas but the country's economy was a dismal underperformer and the average income of Soviet citizens was a fraction of the French or German mean, although those two countries had to import virtually all of their oil. Modernizing (an adjective I prefer to developing) oil-rich nations in general, and OPEC nations in particular, provide even better examples of this reality. Except for Nigeria (190 million people in 2017) and Iran (80 million), OPEC nations have very small or relatively small populations, and since the early 1970s they have benefited (albeit, as just explained in the previous section, in a highly fluctuating manner) from enormous transfers of wealth from oil-importing countries. In 2012, the year of record profits, the current account balance of OPEC countries was nearly half a trillion dollars in the black – but in 2015 it was $100 billion in the red, as Saudi Arabia went from +$165 billion ($5,650/capita) to -$41 billion. The past periods of fabulous earnings have visibly transformed all Middle Eastern oil producers: but judging the progress by the number of new skyscrapers, giant airports and shopping centers would be misleading as economic and social advances have not been commensurate with the new riches.

OIL, HUMAN DEVELOPMENT, FREEDOM AND CORRUPTION

One of the most revealing international comparisons is the Human Development Index (HDI) that is made up of three major components: life expectancy at birth; adult literacy rate and combined gross enrolment ratio in primary to tertiary education; and GDP per capita expressed in terms of purchasing power parity.

OIL, HUMAN DEVELOPMENT, FREEDOM AND CORRUPTION (cont.)

This simple shortcut serves well as an indicator of a nation's relative achievements and it reveals that none of the oil-rich Middle Eastern nations is performing well. There are 188 nations listed in the 2015 edition of the *Human Development Report,* with ten countries at the top (including Norway, Australia, the US, Canada and New Zealand). But Saudi Arabia is ranked 38th, United Arab Emirates 41st, tiny, super-rich Kuwait 48th and Iran 69th (and Nigeria occupies 152nd place, even behind 149th Angola).

Oil-rich Middle Eastern under-performers

country	HDI rank	GDP minus HDI rank	political freedom index	corruption perception index	corruption perception rank
Qatar	40	-13	6	6.0	32
United Arab Emirates	41	-18	6	6.2	31
Kuwait	44	-11	4	4.8	46
Oman	71	-30	6	5.4	39
Saudi Arabia	77	-33	7	3.3	70
Iran	99	-29	7	2.7	105

HDI rank among 177 ranked countries. Political freedom index: 6-7 not free; 4-5 partially free.
Corruption perception rank among 163 ranked countries.

Figure 6 Oil-rich Middle Eastern underperformers

Even more telling is the difference between the rank of the average per capita gross national income (GNI) of a country and its HDI rank: positive numbers identify the nations whose state of development is higher than expected when judged solely by their GDP; negative numbers apply when a nation's HDI lags behind its GNI. In this respect, all oil-rich Persian Gulf nations have been dismal underperformers: not only do they have negative scores, but these scores are among the highest worldwide, indicating that no other group of countries has used its riches so unwisely (see figure 6). Rankings by the political freedom index put five Persian Gulf oil producers (Saudi Arabia, Iran, Iraq, UAE, Qatar, Oman) as well as Libya, Algeria and Venezuela into the bottom, not free, category and Kuwait and Nigeria into the lower ranks of the partly free group. Another revealing set of comparisons concerns corruption. Long ago, Pablo Pérez Alfonso, a founder of OPEC, spoke of oil as a curse, bringing waste, corruption and excessive consumption, and

**OIL, HUMAN DEVELOPMENT, FREEDOM
AND CORRUPTION (cont.)**

his conclusion (oil-rich Norway aside) has been amply confirmed with every new entrant, most recently Angola (joined OPEC in 2007) and Equatorial Guinea (oil discovered in 1995, allowed to join OPEC in 2017). Transparency International's Corruption Perception Index for the year 2016 ranges from the cleanest, Denmark (90), to hopeless Somalia (10). Libya and Iraq were at near-Somalia levels, Angola got a dismal 18, Nigeria got 28 and Russia joins this unenviable oil-rich group with a corruption index of 29, the same as Iran, while Saudi Arabia got 36 (see figure 6).

The evidence is clear: oil-rich countries in general, and the Middle Eastern ones in particular, have not used their considerable wealth to build more equitable and less corrupt societies with a higher quality of life. In fact, the opposite is true as they have too often embodied (and to an exceptionally high degree) many negatives that prevent real modernization of their societies: record-breaking skyscrapers and gargantuan airports do not make up for such fundamental deficits.

And it does not get any better once we turn to the stability of governments and civil institutions and long-term prospects for security. There are only two non-Western oil-rich countries (Qatar and United Arab Emirates) whose politics and stability are not matters of chronic concern and anxious speculation. Worries about long-term stability in most OPEC countries, and particularly in Saudi Arabia, Iran and Nigeria, have, for decades, generated vast speculative literature and more warnings and catastrophic scenarios are sure to come. Perhaps the most useful fact to keep in mind is that most of these writings are produced by people who understand neither Arabic nor Farsi and whose knowledge of the culture steeped in Islam is limited to repeating such often misunderstood terms as *jihād* or *fatwa*.

Any assessment of the Middle East's largest oil producer must take into account the secretive nature of decision making within Saudi Arabia's extensive ruling family and the complexity of the country's traditions and its slowly unfolding reforms.

Unfortunately, we are often offered caricatures rather than revealing portraits of the country. Most notably, the Saudi royal family has been portrayed for decades as unstable, insecure and out of touch, and many commentators have explicitly predicted its imminent (and violent) demise when the corrupt and incompetent princes will be swept away by religious zealots or by impatient reformers. Yet, somehow – and with only relatively small incremental concessions toward broader democratic rights (such as finally allowing women to vote in municipal elections in 2015) – the family has remained in control. Of course, this simple observation is not proof of the family's real and lasting stability, merely a reminder that prevailing Western analyses have been repeatedly wrong.

The analytical and interpretive task is no easier as far as Iranian affairs are concerned. Outsiders have to reckon with the complex dynamics of the now decades-long struggle for influence and power among the ruling fundamentalists. Nationalist zealots have been conveniently strengthened not only by periodically high oil earnings but also by the US invasion of Iraq that opened the way for now pervasive Iranian support of the *shī'ī* leadership in Baghdad. As a result, some Western commentators have been portraying Iran as an almost superpower-like actor whose expansive designs for a new *shī'ī* dominated Middle East (ominously reinforced by the country's quest for nuclear weapons) will, at a minimum, destabilize the entire region and may even lead to a global conflict. But the country has also had some more pragmatic leaders who may be supported (on a few occasions openly by street demonstrations) by a large proportion of the country's young (but tightly controlled) population but who are unlikely to gain control of the fundamentalist state.

Matters do not get any better when attention is turned to the violence and extraordinary level of corruption and political tension in Africa's most populous state, Nigeria. Oil production in the Niger Delta has been repeatedly disrupted by rebels who blow up pipelines and oil terminals, and kidnap oil workers and executives, even as chronic discord between the federal and state governments is being exacerbated by rising income inequalities, persistence of extraordinary levels of corruption, radicalization of the Muslim North and, since 2010, violent clashes with Boko Haram, a terrorist outfit operating mostly in the state of Borno. Contrary to the direst predictions, Nigeria may not dissolve anytime soon in a new civil war, but its seemingly intractable problems guarantee the continuation of protracted abysmal economic underperformance and endless political, social and security crises.

And the litany of concerns does not end with these great oil players. Algeria has been through a long and brutal civil war as the secular government fought the fundamentalists, and the stability of its regime remains uncertain. For a short time the prospect was more promising in Libya after its leader had publicly forsaken decades of erratic and violent ways (bombing a Pan Am flight in 1988, supporting terrorist movements abroad, developing nuclear weapons) but this new-found moderation was short-lived: by 2011 Gaddāfī was dead and Libya split into warring factions that were unable to agree on reconstituting an effective and truly national government (although this has not prevented considerable recovery of oil output in 2016 and 2017).

Oil's impact on politics and policies is also ever-present in affluent countries whose prosperity is underpinned by large-scale oil imports, and its expressions range from questionable attitudes toward oil-exporting nations to the use of oil imports as a reason for not just advocating but heavily subsidizing alternative energy sources. Attitudes toward oil-producing nations find their most extreme expression with respect to Middle Eastern countries. Among the Western political elites these feelings have ranged from kowtowing to unsavory rulers and eagerly selling them arms all the way to calls for the US to rethink (at a minimum) its ties with what many commentators see as terrorism-breeding, fundamentalist, treacherous and family-ruled Saudi Arabia. Many Western politicians and activists argue the best long-term solution to cutting the addiction to oil imports from the chronically unstable Middle East is to achieve energy independence.

During the late 1970s this led to a very expensive commitment to a massive development of oil from the Rocky Mountains oil shales (promptly aborted), and subsequently this quest led to an uncritical embrace of heavily subsidized, environmentally unfriendly and barely net energy-positive corn-derived ethanol. From a purely economic standpoint it is counterproductive to divert limited resources to endeavors that would supply a more expensive substitute and in ethanol's case also one whose production would yield only a small net energy return. Although the chances of any lasting embargo on oil exports from countries whose very survival depends on them are highly unlikely (Gaddāfī's Libya or Khomeini's Iran were equally faithful suppliers of oil to the West as its supposedly great ally, Saudi Arabia), a move toward reduced dependence on imports makes economic and strategic sense. A greater level of domestic energy self-sufficiency would improve the trade balance and, although oil-producing countries might always be willing to supply the fuel, the certainty of shipments could be interrupted by a hostile power. Such a scenario has become more likely following Iranian boasts about having the ability to close the Strait of Hormuz to tanker traffic and Chinese claims to sovereignty over nearly all of the South China Sea.

For decades, American dreams of energy independence remained just that, but that elusive goal has been, rather suddenly and unexpectedly, brought much closer to reality by rapid adoption of horizontal drilling and hydraulic fracturing of shales, an innovative extraction method whose details will be explained in the fourth chapter. Fracking, as the practice is commonly called, has had a stunning impact as it has made the US, once again, the world's leading producer of hydrocarbons.

OIL AS *CASUS BELLI*

Oil's strategic role has been consistently overplayed by some careless historians. The most notorious example of these exaggerated claims is that Japan attacked Pearl Harbor in 1941 because in January 1940 the Roosevelt administration abrogated the 1911 Treaty of Commerce and Navigation, in July 1940 stopped licensing for exports of aviation gasoline and in September 1940 added a ban on exporting scrap iron and steel. Apologists for Japan thus argue that this forced Japan to attack the US in order to secure its access to Indonesian and Burmese oilfields. But claiming that is to ignore the fact that Japan began its conquest of Manchuria in 1933, culminating in an attack on China proper in 1937, and that if this aggression against China had been abandoned the country could have maintained free access to all imports: clearly, attacking Pearl Harbor was a self-inflicted blunder. And while it is true that Hitler tried to capture the rich Baku oilfields after the invasion of the USSR, it is obvious that Germany's serial aggression against countries such as Czechoslovakia, Poland, France, the UK, Yugoslavia, Greece and the USSR was not motivated by controlling foreign oil production.

In contrast, indirect foreign interventions in Middle Eastern countries (arms sales, military training, generous economic aid) have aimed either at stabilizing or subverting governments in the oil-rich region. Their most obvious manifestation during the Cold War was the toppling of Mossadegh's government in Iran in 1953; the sales (or simply transfers) of Soviet arms to Egypt, Syria, Libya and Iraq; the concurrent American arms shipments to Iran (before 1979), Saudi Arabia and the Gulf states; and the Western support of Iraq during its long war with Iran (1980–1988). And, of course, the Gulf War (1991) and the US invasion of Iraq in 2003 have often been portrayed as purely oil wars.

Saddām Hussein's occupation of Kuwait in August 1990 doubled Iraq's crude oil reserves (to about 20% of the world total) and it also directly threatened the nearby supergiant Saudi oilfields and hence the survival of the monarchy that controls a quarter of the world's oil reserves. The massive anti-Saddām coalition and half a million troops engaged in Operation Desert Storm in 1991 could thus be seen as a perfect example of an oil-driven war. But other concerns were also at play: Hussein's quest for nuclear weapons with which the country could dominate and destabilize the entire region, and the risk of another Iraqi–Iranian or Arab–Israeli war. And if the control of oil was the primary objective

OIL AS *CASUS BELLI* (cont.)

of the 1991 Gulf War, why then were the victorious armies not ordered to occupy at least Iraq's southern oilfields?

Similarly, more complex considerations were behind the conquest of Iraq in March 2003. The two most important factors were a decade-long refusal of the Iraqi regime to comply with numerous UN resolutions, and the traumatic impact of 9/11 attacks on US foreign policy. Both of these factors led to a shift among the international community from trying to isolate a hostile regime in Baghdad to the pre-emption of a possible new attack, a fear based, as became clear later, on mistaken assumption about Iraq's advances in producing weapons of mass destruction. At the same time, there was an implied grand strategic objective of eventually having an elected government in a pivotal state in the Middle East that might serve as a powerful and stabilizing political example in a very unsettled region and be a mighty counterweight to any radicalizing tendencies: that this has not worked out does not invalidate the original intent.

In any case, what many commentators saw simplistically as a clear-cut case of oil-driven war has been anything but. The invasion (2003) and occupation of Iraq (the last regular US forces were withdrawn in 2011) exacted high human and economic tolls (conservatively put at hundreds of millions of dollars) and once the country's oil production began to recover the US was not its primary beneficiary. In 2012, a year after American withdrawal, 72% of Iraqi crude oil exports went to East Asia (mostly China) and the EU, and by 2015 that share was up to 90%. Obviously, the US has never had any existential need for Iraqi oil (and does so even less now as it has regained the status of the world's largest oil producer), but judging by the destination of Iraqi oil exports should we then conclude that the US fought in Iraq to benefit China and the EU?

But there is no doubt about the importance of oil for modern armies. World War I was dominated by railways, cavalry, horse-drawn wagons and guns and forced marches. During World War II German and Soviet armies still deployed large numbers of horses, but it was the first largely mechanized conflict relying on trucks, tanks and planes. Fuel demands rose afterwards with the development of better armed and more powerful tanks and with the introduction of jet aircraft. America's 60t M1A1 Abrams battle tank consumes kerosene at no less than 400 l/100km (for comparison, the 2017 Mercedes model S600 needs 21 l/100km in city driving, and the 2017 Honda Civic needs 10 l/100km). Kerosene requirements of supersonic combat aircraft are so high that no extended mission can be flown without in-flight refueling from tanker planes. Not surprisingly, the US Department of Defense has by far the highest oil consumption of all government agencies: in 2015 its demand accounted for about roughly 90% of all refined fuels bought by the government. But, again, a caveat is in order: oil supremacy is not a decisive factor in asymmetrical conflicts, a reality amply demonstrated in Vietnam, by the plane-borne attacks of September 11, 2001 and by numerous suicide bombings (above all in Iraq and Pakistan) during the subsequent years.

Oil and the environment

Perhaps the most newsworthy environmental impacts of the oil industry are the periodic accidents in which giant tankers spill large volumes of oil into the sea and onto beaches, resulting in the long-lasting pollution of beaches or rocky shores and in the highly publicized mass mortalities of sea birds. Less noticeable is the contamination of zooplankton and the persistent presence of oil in anoxic sediments that has a long-term influence on benthic invertebrates. The worst tanker accidents have been those of the *Atlantic Empress* that spilled 287,000t off Tobago in 1979 and the *ABT Summer* that released 260,000t off Angola in 1991. Both of these mishaps took place far offshore and hence they received much less attention than the world's third and fourth largest record spills, the *Castillo de Bellver* that released 253,000t of crude off South Africa's Saldanha Bay in 1983, and the *Amoco Cadiz*, much of whose cargo of 223,000t of light crude ended up on the beaches of Brittany in 1978. So far, the largest tanker oil spill in the twenty-first century was the loss of 63,000t of oil from the *Prestige*, a Greek single-hull ship in Galician waters in November 2002.

Studies of oil spill causes show that groundings, collisions and hull failures have been (in that order) the main reasons for these mishaps. The good news is that the frequency of both large and small spills has been constantly declining since the 1970s and that the aggregate quantity spilled annually from ships has been for many years well below the amount of oil reaching the sea from natural seeps. The bad news is that long-term studies of oil spill sites have shown unexpected persistence of toxic subsurface oil and chronic sublethal exposure with lasting effects on wildlife. Much of this new understanding was gained by follow-up studies of the most notable North American tanker spill, that of the *Exxon Valdez* in Alaska's Prince William Sound on March 24, 1989. The grounded ship released only 37,000t of oil but the spill killed perhaps as many as 270,000 water birds and it left many long-lasting effects on marine biota.

Exxon paid out about $2 billion for the oil clean-up and another billion to the state of Alaska, and the costs of restoring the waters and (at least superficially) the rocky shores and beaches of Prince William Sound were thus internalized to a degree unprecedented in previous oil spill accidents. And much higher fines and other compensation payments (totaling around $50 billion by 2017) were paid by British Petroleum after the wellhead blowout on *Deepwater Horizon*, a semisubmersible drilling rig in the Gulf of Mexico, led to the loss of as much as 620,000t between April and July 2010. In contrast, the Mexican IXTOC 1 well in Bahia de Campeche spilled perhaps as much as 1.4Mt in 1979–1980 without any penalties being paid by Pemex. Fortunately, most of the small-scale spills of crude oil or refined products from coastal or river-going vessels do not overwhelm the natural processes of evaporation, emulsification, sinking, auto-oxidation and, most importantly, microbial oxidation, that limit their impact on surface waters.

Risks of major spills have been lowered by better regulation: all oil tankers of 600 tonnes deadweight and above delivered after July 6, 1996 must have double hulls and double bottoms. What has not changed is the flagging and crewing of oil tankers. Most of the world's tankers (and other freight vessels) fly a flag of convenience, which means that their ownership and control has nothing to do with the country of registration. Such registrations, now offered by nearly thirty countries (Liberia, Taiwan, Honduras, Belize, Panama, Malta and Spain are the leading flag-of-convenience providers, but the list also includes landlocked Bolivia and Mongolia) provide cover for substandard practices and for evading legal responsibility for oil spills. And the crews are overwhelmingly Asian: the Philippines and India supply nearly half of all officers and 90% of lower level crews.

The Kuwaiti well fires of 1991 were perhaps the most prominently reported environmental catastrophe involving the combustion of crude oil. More than 700 oil and gas wells were set ablaze (it took nine months to extinguish them) and because the very small particles generated by oil combustion can stay aloft for weeks they were carried far downwind: only ten days after Iraqi troops set fire to Kuwaiti oil wells in late February 1991 soot particles from these fires were identified in Hawaii. In subsequent months, solar radiation received at the ground was reduced over an area that extended from Libya to Pakistan, and from Yemen to Kazakhstan. But oil combustion has a much more important effect environmentally, and healthwise, because of its generation of the three key precursors of photochemical smog, carbon monoxide, volatile organic carbohydrates (VOC) and nitrogen oxides (NO_x).

Photochemical smog was first observed in Los Angeles in the 1940s and its origins were soon traced primarily to automotive emissions. As car use progressed around the world all major urban areas began to experience seasonal (Toronto, Paris) or near-permanent (Bangkok, Cairo) levels of smog, whose effects range from impaired health (eye irritation, lung problems) to damage to materials, crops and coniferous trees. A recent epidemiological study in California also demonstrated that the lung function of children living within 500m of a freeway was seriously impaired and that this adverse effect (independent of overall regional air quality) could result in significant lung capacity deficits later in life. Extreme smog levels now experienced in Beijing, New Delhi and other major Chinese and Indian cities arise from the combination of automotive traffic and large-scale combustion of coal in electricity-generating plants and are made worse by periodic temperature inversions that limit the depth of the mixing layer and keep the pollutants near the ground.

Introduction of three-way catalytic converters (reducing emissions of CO, VOC and NO_x) helped to limit smog levels but their use had to be preceded by production of unleaded fuels in order to avoid the poisoning of the platinum catalyst. By that time decades of leaded gasoline consumption had created high levels of lead contamination in all urban areas with high traffic density. Lead's phase-out began in the US in 1975 and it was completed by 1990. Methyl tertiary butyl ether (MTBE, produced from isobutylene and methanol) became the most common additive to boost octane rating and to prevent engine knock, and starting in 1995 it made up as much as 15% of the reformulated gasoline designed to limit air pollution. But because MTBE (a potential but not officially listed human carcinogen) is easily miscible with water and leaks had contaminated many water wells, a switch to ethanol began in 2003 and MTBE as a fuel additive is banned in the US.

Combustion of refined fuels generates less CO_2 per unit of released energy than does the combustion of coal, but in aggregate they became the world's largest source of carbon from burning fossil fuels in 1968 when they accounted for about 43% of the total. The share of carbon from liquid fuels rose to nearly 50% by 1974 but during the late 1980s it was basically the same as for coal combustion only to pull ahead once more during the 1990s – but by 2004 coal (thanks to China's rapidly expanding extraction) was once again the largest source of fossil carbon. In 2015 liquid fuels contributed about 34% of all carbon from fossil fuel combustion, while emissions from coal accounted for about 41%. Unlike solid fuels, whose emissions now come, almost without exception, from stationary sources (potentially controllable by the sequestration of CO_2), the bulk of the carbon emissions from liquid products comes from the transportation sector and the only possible control is to prevent their generation. Combustion of every liter of gasoline releases about 2.3kg of CO_2 and the rate for diesel fuel is 2.6kg/l.

All of these economic, strategic, health and environmental burdens should be considered in any serious attempts at finding the real cost of oil – but these efforts are exceedingly complicated because of the many assumptions, approximations and uncertainties that are required to quantify externalities (What is the cost of a smog-induced asthma attack?), to answer counterfactual questions (How much would the US Department of Defense save if the Middle East contained no oil, but was still full of Muslim *jihādis* bent on attacking the US?), and to set the analytical boundaries (Should the entire cost of urban sprawl be charged to gasoline?).

Obviously, these challenges have no definite solutions and hence the estimates of the real price of crude oil or gasoline can end up with totals only somewhat higher than the prevailing price or costs that are an order of magnitude above the current price. An excellent example in the latter category is the study of the real cost of US gasoline completed by the International Center for Technology Assessment in 1998 when Americans were paying just over $1/gallon. Inclusion of tax and program subsidies to oil companies, protection (mainly military) subsidies, environmental, health and social costs (ranging from air pollution to urban sprawl) and other outlays (from travel delays due to road congestion to uncompensated damages due to car accidents and subsidized parking) raised the real cost to between $5.60 and $15.14/gallon. A 2015 study concluded that the cost of atmospheric emissions alone should raise the price of American gasoline by $3.80/gallon.

I am very much in favor of more realistic cost estimates but I must also note that these exercises have many methodological problems and are inherently biased as they do not consider the many benefits arising from the use of these subsidized fuels. Different assumptions and non-uniform analytical boundaries result in incompatible conclusions: do the US oil companies enjoy unconscionable subsidies or are their benefits relatively minor? Does the production and use of automobiles add up to a net benefit or a net burden in modern economies? Ubiquitous examples of unaccounted benefits range from lives saved by rapid transfers of patients to hospitals by ambulances, time saved by flying as opposed to taking trains, or better air quality compared to burning wood or coal. No single answer is possible but the key conclusion stands: prices paid for refined oil products certainly do not reflect their real cost to society.

2
What oil is and how it was formed

Most people are fairly familiar with at least one of the many economic, social and political consequences of humanity's dependence on oil that were reviewed in the first chapter. In contrast, most people know little or nothing about the intricate and complex origins of crude oil and about the actual composition of this valuable commodity that has such a central role in our lives (crude oil's essential properties and measures are summarized in Appendix B).

The generic term 'oil' designates two very different kinds of compounds that are liquid at ambient temperature and normal atmospheric pressure. Crude oils (using the proper qualifying adjective and a plural whose importance will be explained shortly) that oil companies extract from the topmost layer of the Earth's crust belong, together with other mineral substances whose composition is dominated by hydrogen and carbon, to a class of compounds called hydrocarbons: these can be gaseous (natural gases), liquid, or solid (bitumens, tars). Oils that abound in the biosphere and that form an important part of our diet (as plant oils extracted from seeds and nuts) or enhance our drinks and perfumes (delicate essential oils in herbs or the rinds of citrus fruit) are not, despite their similar make-up, classed as hydrocarbons, while the reverse term, carbohydrates, is reserved for organic molecules whose carbon and hydrogen atoms form simple sugars (mono- and disaccharides) whose polymers yield starches and cellulose, the biosphere's most abundant organic materials.

WHAT'S IN A NAME?

نفت (*naft*), an ancient Persian word for a dark liquid coming from the ground, was adopted in classical Greek, but most European languages eventually accepted variations of the Greek rock oil, πετρέλαιο (*petrélaio*) – English petroleum, French *pétrole*, Italian *petrolio* – or its literal translations, such as German *Erdöl* or Dutch *aardolie*. Because petroleum has to be processed (refined) to produce a variety of liquid fuels that are suited for specific uses, an accurate generic term for the substance, and a common American usage, is simply crude oil (German *Rohöl*, French *pétrole brut*). In modern Western usage, the term naphtha has contracted to designate a specific range of refined crude oil products (see chapter 4).

The ancient Persian word was retained in its original broad meaning as the Arabic النفط (*al-naft*) and it also migrated into Slavic languages: Russian нефть (*neft'*), Czech *nafta*. However, modern Arabic now habitually uses the Greek import, as newspapers write about بترول (*bitrūl*). Chinese and Japanese use the same two characters 原油 for the crude oil; their standard (*Putonghua*) Chinese pronunciation is *yuán yóu* and Japanese pronunciation is *genyu*. Petroleum is translated into Chinese and Japanese literally as rock oil, 石油 , pronounced as *shí yóu* in Chinese and *sekiyu* in Japanese.

A plural rather than a collective singular is a much more accurate designation of crude oils because they encompass substances of considerable heterogeneity and vary widely in their appearance, composition, viscosity, flammability, quality and hence in economic usefulness, and command commensurately different prices. Moreover, these differences are seen not only among different oilfields but even within sections of a single large oil-bearing rock formation. And as some crude oils are extremely viscous and hardly flow, and as there are enormous volumes of oil locked in tiny droplets interspersed in sands and shales, it is not their liquid state but the shared presence of several series of hydrocarbon compounds that provides the rationale to subsume the extended continuum of oils under a homogenizing singular.

The origins of oil are also sufficiently complex to leave room for two diametrically opposed theories of hydrocarbon formation: the dominant theory sees oils and gases as products of complex, gradual transformations of ancient accumulations of dead organic matter, and the Russian–Ukrainian school of petroleum geology that sees hydrocarbons as abiogenic products, formed under high pressures and temperatures deep in the Earth's mantle from which they rise to be trapped in porous structures near the planet's surface. I will close the chapter by reviewing some essential concepts and processes of petroleum geology such as depositional systems, reservoir rocks and oil migrations and relating these settings and events to the major petroleum provinces and largest oilfields.

Composition of crude oils

Most adults are familiar with the color (or lack of it) and characteristic smell of common refined oil products (gasoline, diesel oil, kerosene, lubricating oils, asphalt) but few have seen crude oils as they come from the ground or as they are carried by tankers and pipelines into refineries. The appearance of crude oils ranges from light, gasoline-like, highly mobile liquids to heavier reddish-brown fluids to highly viscous, tarry black materials. Crude oils are not chemical compounds (hence it is impossible to write down their formulas); they are complex mixtures of scores of hydrocarbons and other minor ingredients. Ultimate elemental analysis of crude oils shows carbon accounting for about 85% (83–87%) of their mass, and hydrogen for 13% (11–15%), with hydrogen-to-carbon ratios around 1.8, compared to about 0.8 for bituminous coals and 4 for methane.

The smallest hydrocarbon molecule present in crude oils is gaseous methane, the largest one has more than eighty carbon atoms. Three series of hydrocarbons (whose composition differs by a predictable number of carbon and hydrogen atoms according to a generic formula) dominate the composition of crude oils (see figure 7). Scientific names of these series are alkanes, cycloalkanes, and arenes but in the oil industry they are known as paraffins, cycloparaffins (or naphthenes) and aromatics. Alkanes are the second most abundant homologous series found in crude oils, making up about a quarter of the total mass. Their common name, paraffins (from Latin *parum affinis*, of slight affinity), refers to their inert nature: they do not react with either strong acids or alkaline (or oxygenating) compounds.

They are either straight-chain (normal) or branched-chain molecules; the first group has some sixty members, the other runs theoretically into millions. The two lightest straight-chained alkanes, methane and ethane, are gases at atmospheric pressure. Propane and butane are also gases but are easily compressible to liquids (hence known as liquid petroleum gases, LPGs). Chains with five (pentane) to sixteen carbons are liquids, and the remainder are solids. Pentane, hexane and heptane are normally the most abundant alkanes in crude oils. Natural gas liquids (NGLs) are hydrocarbons of low molecular weight that are dissolved in natural gas (be it associated with oil or in non-associated reservoirs) and that get separated from the gas at special processing facilities: they include ethane, propane, butane and isobutene (LPGs) and gas condensate, that is, alkanes containing two to eight carbon atoms (C_2H_6 to C_8H_{18}). Production of NGL is sometimes reported separately from the extraction of crude oil. *Oil & Gas Journal*, one of the industry's leading publications, excludes NGLs from its tabulations and hence its totals of global crude oil production are always lower than all other commonly used statistical series that include NGLs, including those of the US government, United Nations and British Petroleum.

alkanes

pentane

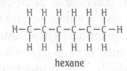

hexane

cycloalkanes

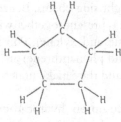

cyclopentane

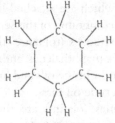

cyclohexane

arenes

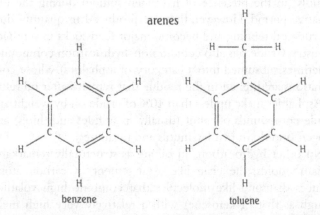

benzene

toluene

Figure 7 Structures of the most common liquid alkanes, cycloalkanes and arenes

Cycloalkanes (naphthenes in the oil industry) are the most abundant compounds in crude oil (typically half of the weight), with methylcyclopentane and methylcyclohexane present in the greatest quantities. These saturated hydrocarbons have carbon atoms joined in rings of five (cyclopentane) or six (cyclohexane) atoms which tend to fuse into polycyclic molecules in heavier fractions: bicyclic naphthenes are common in kerosene, tetracyclic and pentacyclic compounds are present in lubricating oil. Arenes (aromatics) are unsaturated, highly reactive liquids named after the members with pleasant odors that share at least one benzene ring to which are attached long, straight side chains. Benzene is the first compound of this series and it is present together with its alkyl derivatives (toluene, ethylbenzene and xylene). Polycyclic aromatics (naphthalenes, anthracene and phenanthrene) are most common in heavy oils and lubricants and they make up less than 20% of crude oil mass.

Alkenes (olefins) are the most common hydrocarbons in organic matter (present as oils in both plant and animal tissues), but only their traces are found in some crude oils because they are easily reduced to alkanes (in the presence of hydrogen) or to thiols (in the presence of hydrogen sulfide) during the oil's formative period. However, they are produced in quantity during crude oil refining and become major feedstocks in synthetic chemistry. Crude oils also contain non-hydrocarbon compounds (sometimes subsumed into a category of asphaltics) whose concentrations are highest in the residue that remains after oil refining. Asphaltics make up less than 10% of crude oil by weight and include compounds of sulfur (usually in sulfides and thiols) and nitrogen (mostly in pyroles, indols and pyridines).

No other hydrocarbons in oil are as structurally remarkable as diamondoids; the chair-like arrangement of carbon atoms produces diamond-like molecules that combine high volatility (as high as that of kerosene) with a relatively very high melting point (higher than that of tin). This extraordinary thermal stability means that their concentrations increase with increasing temperature and that their presence can be used to narrow down the maximum temperatures compatible with the generation of oil hydrocarbons in source rocks and to estimate the degree of oil destruction by natural cracking processes.

Sulfur (ultimately derived from sulfur bonds present in proteins of ancient organic matter) is the most common undesirable contaminant of crude oils because its combustion generates sulfur dioxide, a leading precursor of acid rain. So-called 'sour' oils have more than 2% of sulfur (which means that in residual oils and in asphalt the levels are commonly above 5%), while 'sweet' crude oils have less than 0.5%, with some of them (from Nigeria, Australia and Indonesia) having less than 0.05% S. In addition to dominant hydrocarbon homologs and non-hydrocarbon compounds, some crude oils also have relatively high (but in absolute terms only a tiny fraction of 1%) traces of aluminum and heavy metals, including chromium, copper, lead, nickel and vanadium.

Straight distillation of lighter oils will produce larger shares of lighter fractions: naphtha can be above 20% for Nigerian light oil and kerosene almost another 20%, with the residue accounting for just a third of total weight; in contrast, distillation of a heavy Kuwaiti oil will yield less than 25% of naphtha and kerosene and leave nearly two-thirds as a residue. Light oils are naturally preferred by markets that are dominated by gasoline and kerosene demand because their refining needs less catalytic cracking (see chapter 4) to produce the needed mix of products. Oils with high paraffin content have elevated pour points. While light crude oils can flow at temperatures as low as -50°C (and commonly at less than -20°C), some high-paraffin crudes will gel even at 40°C, and even those with pour points above freezing point may have to be heated before they can be transported in pipelines in cold climates, or special additives must be used to lower their viscosity. Daqing oil, from China's largest field, is an example of very waxy crude: it contains about 26% wax by weight and its gelling point is 32°C.

Because the liquid hydrocarbons forming crude oils have very similar specific heat content (or energy density, that is, energy released by the combustion of a unit mass of a substance), the total energy content of different crude oils remains fairly uniform, ranging between 42–44 megajoules per kilogram (MJ/kg), and international energy statistics usually use 42MJ/kg (or 42GJ/t) as the typical value when converting fossil fuels to a common energy denominator. This means that the energy density of crude oils is about 50% higher than that of the best anthracite coals (29–30MJ/kg), about twice as high as that of common steam coals (20–24MJ/kg) used for electricity generation, roughly 2.5 times as high as that of air-dry wood (16–18MJ/kg) and four times higher than that of the low-ranking lignites (brown coals). Low density of gases means that a cubic meter of natural gas will contain about as much energy as a liter (1/1000 of m^3) of crude oil.

Oil's high energy density is perhaps the fuel's most important advantage when compared to coal but there are other notable reasons why liquid fuels have been preferred to solids and why the post-WWII global energy transition from coal to oil has proceeded so rapidly. The already noted higher hydrogen/carbon (H/C) ratio of crude oil means the combustion of refined products generates 20–25% less carbon dioxide, the most important greenhouse gas, per unit of energy than coal. Although the sulfur content of some crude oils is high, refined products contain much less sulfur than coal and their combustion yields much less sulfur dioxide, the gas most responsible for acidifying deposition (acid rain). And, unlike the burning of coal, combustion of liquid fuels produces only trace amounts of particulate matter.

Other obvious advantages are the ease of long-distance transport: unlike coal, crude oil and refined products can be transported inexpensively, conveniently and very safely by pipelines;

and worldwide, very cheaply but with higher risks, by tankers. Unlike coal, liquid fuels are also conveniently stored in large above-ground tanks, underground reservoirs or natural caverns filled by pumping. And refined products have a wide range of uses: they can heat homes and industries, generate electricity, power all modes of land, water and air transportation; some oil fractions also have important non-fuel uses as chemical feedstocks and as lubricants and paving materials. Finally, crude oil reaches the surface either because of naturally high reservoir pressure or because of a mechanized artificial lift (mostly by pumping) and its production does not require any dangerous underground work; similarly, refining and distribution processes are highly automated, limiting risky occupational exposures.

Origins of oil

Standard geological consensus sees fossil fuels as organic mineral-oids formed by the accumulation, burial and transformation of ancient dead biomass, the remnants of terrestrial and aquatic plants and heterotrophic organisms. They are found in the Earth's crust in all three forms of matter, as solids (coals, peats), liquids (crude oils) and natural gases. As already explained, plurals are needed to convey the considerable heterogeneity of their chemical composition and physical properties. All coals were formed through the accumulation and transformation of plant mass (phytomass), the product of enzymatically mediated conversion of solar (radiant or electromagnetic) energy into chemical energy of new plant tissues. These conversions took place in environments akin to today's peat swamp forests of South-East Asia. Often, exquisitely preserved imprints of leaves and fossilized twigs, branches and trunks offer abundant testimonies of this origin.

Photosynthetic conversion is ancient: organic carbon in Archaean sediments puts the first prokaryotic photosynthesizers

3.8 billion years ago, but massive deposits of coal date only from the time when high rates of photosynthesis by large terrestrial plants left behind enormous deposits of dead organic matter. By far the largest coal resources are of Palaeozoic origin (544–245 million years ago, mostly from the Carboniferous period, 359–299 million years ago), and most of the rest from the middle and upper Mesozoic era (Jurassic and Cretaceous period) and from the oldest Cenozoic era (from the Paleocene period). Only the poor quality lignites and peat are the products of the Quaternary period whose oldest sediments were laid down less than 1.8 million years ago (see figure 8).

Oil's origins are not as easy to trace as those of immobile coal imprinted with visible plant signatures. Oil's mobility means that it is usually found in places where it was not formed and the susceptibility of hydrocarbons to alteration by chemical and physical means led to speculations that it originated from the transformation of coal or by polymerization of gases from basal rocks. There is no doubt that liquid and gaseous hydrocarbons can be formed directly from their constituent atoms by inorganic chemical reactions, specifically by polymerization of methane precursors, but the best isotopic evidence suggests these abiogenic processes do not produce globally significant volumes of crude oils.

Modern consensus among petroleum geologists and geochemists is that oils of inorganic origin are commercially unimportant and that crude oils are derived from dead biomass, from organic compounds formed mostly by monocellular phytoplankton (dominated by cyanobacteria and diatoms) and zooplankton (above all by foraminifera) as well as by higher aquatic plants (algae), invertebrates and fish. In addition, terrestrial organic matter, carried by rivers into the ocean, has also been an important contributor of sedimented biomass, and many oil formations had their origin in rich lake (lacustrine) environments. Only a small fraction of oil was formed directly by accumulation of biogenic hydrocarbons and their subsequent transformations.

Era	Period	Beginning (million years before present)	Notable life milestones
Cenozoic			
Quaternary	Holocene		Civilization
	Pleistocene	1.8	*Homo sapiens*
Tertiary	Pliocene	5.3	*Australopithecus*
	Miocene	23.8	*Ramapithecus*
	Oligocene	33.7	First elephants
	Eocene	54.8	First horses
	Paleocene	65	First primates
Mesozoic	Cretaceous	144	Flowering plants
	Jurassic	206	First birds
	Triassic	248	First mammals
Paleozoic	Permian	290	Reptiles
	Carboniferous	354	Winged insects
	Devonian	417	First sharks
	Silurian	443	First bony fishes
	Ordovician	490	First vertebrates
	Cambrian	543	Trilobites
Proterozoic		2,500	Oxygen rises
Archean		3,800	Bacteria, archaea

All divisions according to the Geological Society of America.

Figure 8 Geological timescale

The same hydrocarbons that make up crude oil – alkanes, cycloalkanes and arenes – are present in plants, albeit only in trace quantities and in a very limited variety. Remarkably, plant alkanes have only odd numbers of carbon (15–21 for oceanic species, 25–37 for land plants) in their chains. This oddity is retained in immature source rocks but in crude oils, after the plant hydrocarbon chains are randomly broken up, the numbers of odd and even carbon molecules are equal. Terpenes – polymers of isoprene – are another important group of plant hydrocarbons that are commonly encountered as resins (especially in conifers), essential oils (limonene, responsible for citrus scent) and other important compounds, including retinol (vitamin A), and lycopene in tomatoes and carotenes. Some compounds derived from terpenes are considered important biomarkers in crude oil.

Oil appears to be derived mostly from non-hydrocarbon organic molecules that underwent microbial metabolism (bacteriogenesis) and, above all, prolonged thermal decomposition (thermogenesis) after burial in sediments. A recent discovery attests to the ancient origins of this process: abundant nodules of bitumen (dark, inflammable organic matter) and residues of asphaltic pyrobitumen were identified in the black shales of Australia's Pilbara Craton, one of the world's least disturbed granite-greenstone rocks that contain some of the oldest known shales. Their presence indicates that crude oil was generated from organic matter in marine sediments as early as 3.2 billion years ago.

Most of the non-hydrocarbon organic compounds belong to one of the three large categories, carbohydrates (with cellulose, a polymer of the simplest sugar, glucose, being the dominant compound in plants), proteins (forming animal and human muscles) and lipids (fats) composed of fatty acids and glycerol. Organisms also contain such metabolic pigments as chlorophyll (responsible for the green color of bacteria, algae and plants) and hemin (in animals); their porphyrin rings (iron centers) are retained in crude oil, in nickel and vanadium porphyrin complexes that were formed by reactions with haemoglobin and chlorophyll, and they are considered important markers of oil's biogenic origins. The presence of some identical hydrocarbons in plants and crude oils, structural similarities between plant and animal lipids and oil hydrocarbons, and the occurrence of biomarkers confirm oil's organic origins. Many oils also contain fossil spores and pollen.

Most of the world's oil comes not just from sedimentary rocks that were permeated by organic matter but from geological eras known to have exceptionally high photosynthetic productivity, and measurable quantities of oil-like liquid hydrocarbons have been found in a variety of recent marine sediments indicating the continuity of biogenic oil formation. Another notable property arises from the preferential carbon isotope fractionation during photosynthesis: in oil the ratio of two stable isotopes, ^{13}C and the dominant ^{12}C, resembles that of plants (photosynthesis preferentially selects the lighter isotope) and not that of carbonate rocks. And the common presence of nitrogen compounds in crude oils is best explained by their organic pedigree: the element is a critical constituent of amino acids that make up proteins.

Organic matter is transformed into oil by a long, sequential process that starts with accumulation of biomass in sedimentary (marine or lacustrine) environments. Initial microbial aerobic degradation returns a significant part of the sediment carbon to the atmosphere as carbon dioxide. Subsequent anaerobic fermentation by methanogenic and sulfate-reducing bacteria releases methane and hydrogen sulfide. The eventual burial of organic matter in anoxic muds leads to the formation of longer-chained compounds and to the generation of kerogens, complex insoluble mixtures of large organic molecules. The kerogen formed mostly from lipids of both marine and terrestrial origins, with hydrogen-to-carbon ratio <1.25, produces most of the commercially exploited mixtures of crude oil and natural gas.

Kerogen in source rocks (usually shales or limestones) may be as much as 10% of organic matter by weight but contents of 1–2% are typical. The latter suffices to label the rock as a potential candidate for oil generation. With the progressive burial of kerogens comes a rise in temperature (the Earth's temperature normally increases by 25–30°C/km but in many tectonically active regions the rate is much higher) and pressure, and the accumulated organic matter is eventually subjected to thermal degradation (cracking). This process, akin to the production of lighter fuels from heavier fractions in refineries (see chapter 4), breaks up longer-chained molecules and produces lighter compounds. Kerogens change to solid (or almost solid) black or brown macromolecular bitumen whose unmistakable odor betrays its main constituents, heavy hydrocarbons including asphalt and waxes, and the cleavage of bitumen bonds yields complex mixtures of light and heavy hydrocarbon molecules.

Geochemists divide the formation of hydrocarbons from a source rock into three main temperature-dependent stages. Diagenesis (transformation of sediments into sedimentary rocks) is limited to physical and chemical changes that begin at relatively low temperatures of 50–60°C and at depths below 1km. Catagenesis is the principal process of thermal cracking and it is at its most effective within the so-called oil window between 65°C and 150°C, with the ratio of gas/oil formation increasing as the temperature rises with most of oil's complex constituents produced at temperatures between 80°C and 120°C. Metamorphic processes are thermal alterations in environments hotter than 200°C. The oil window can also be defined in terms of depth: where the thermal gradient is high (>5°C) the minimum overburden can be less than 1km, while in formations with low thermal gradient the oil window may extend to below 8km, with optimum oil-producing conditions between about 2.2 and 4.5km.

Source rocks exposed only to lower temperatures are considered to be immature. Those processed extensively at temperatures above 200°C are overmature; crude oil constituents become unstable and dry gas (that is basically pure methane) is the principal product. The rate of oil generation is also affected by the pressure and the presence of heat-tolerant or outright thermophilic bacteria. Hydrogen and oxygen derived from water and surrounding minerals also participate in these transformations, as do catalytically active transition metals. Some hydrocarbon reservoirs yield only negligible volumes of gases, while at the other extreme (where the processing temperatures approached the upper range of the window) there is only natural gas, and during the decades preceding its common commercial use drillers were disappointed to find it and vented or flared nearly all of it.

Most of the oil formed during the past half a billion years was degraded by thermophilic bacteria resident in oil reservoirs and active in temperatures up to about 80°C. These microbes, metabolizing very slowly and acting on geological timescales of millions of years, are able to destroy many of oil's components and produce much denser, heavy oils. These biodegraded oils, rather than the Middle Eastern deposits of hydrocarbon liquids, dominate the world's resources and their largest concentrations are in shallow reservoirs situated on the eastern flank of the American Cordillera. These foreland basins – the Western Canada basin in Alberta and Saskatchewan with its heavy oils and tars, and the Eastern Venezuelan basin (the Orinoco tar belt) – are the world's largest petroleum accumulations (see chapter 5 regarding their exploitation).

Oil generation is clearly a multifactorial and highly protracted process. While the best environment for the formation of crude oils was in the areas that combined generally high photosynthetic productivity with high rates of undisturbed accumulation (deposition) even such ideal conditions would only rarely produce a single massive layer of source rocks because the periods of enhanced photosynthesis (supported by an elevated influx of nutrients) alternated with times of slower growth. Subsequent burial may have been too slow, leading to large aerobic oxidation of the accumulated biomass, and thermal processing could be inadequate or excessive.

Given the abundance of dead organic matter that entered sediments during the past half billion years, and the adequate time that has been available for transforming these remains to liquid fuels, it is clear that the process of oil formation is relatively inefficient. The generation of substantial oil deposits required enormous inputs of primordial organic matter, with no known commercially exploitable oil formations being younger than the time elapsed from the peak of the latest glaciation, and with the oldest ones dating back to the Palaeozoic era. Approximate comparisons with the generation of coal indicate the extraordinary demand of this process: roughly 1,000 times more ancient biomass was needed to transfer a unit of carbon from organic matter to crude oil than was required to preserve it in coal.

PHOTOSYNTHETIC (CARBON) COST OF COAL AND OIL

Coal formation has high carbon preservation rates. Close to 15% of the element is transferred from plants to peat and 75–95% of that carbon ends up as coal. Underground coal mining usually removes about 50% of coal in place and surface mining takes out as much as 90%. The overall carbon recovery factor (the percentage of the element's original presence in phytomass that ends up in marketable fuel) is thus as high as 20% for lignites and as low as 2% for the best anthracites, with rates around 10% being typical for the most commonly mined bituminous coals. Obversely, this means that 5–50 units of carbon locked in ancient plant mass were needed to produce one unit of carbon in coal.

In comparison to coal formation, preservation factors of carbon were lower during the formation of marine and lake sediments (rarely over 10%, often less than 1%), and much lower during the subsequent heating and pressurization of organic sediments. Published data also indicate a much wider range of preservation factors and crude oil also has much lower extraction factors than coal: commonly just 10–20% of all carbon originally present in oil formations ends up on the market. This means that the mean overall recovery factor for crude oil carbon is less than 0.01%. On the average, some 10,000 units of carbon (or as few as about 100 and as many as 300,000 or more) in the initially sequestered biomass were needed to produce a unit of carbon in marketed crude oil. Refining further lowers this recovery factor and a memorable encapsulation of these calculations is that every liter of gasoline (about 740g containing about 640g of carbon) represents some 25t of originally sequestered marine biomass.

Up to this point I have reviewed our understanding of oil's origins based on the prevailing consensus of geologists and geochemists – but I should note that an alternative theory of oil's origins has been available since the early 1950s, an intriguing Russian–Ukrainian hypothesis about the abyssal abiogenic origin of hydrocarbons. This is not yet another example of a deranged Stalinist science (like Lysenkoist genetics): the theory is based on theoretical considerations as well as on extensive field observations and it has been criticized and discussed for decades in more than four thousand papers, book chapters and books. Most importantly, the theory has guided extensive exploratory drilling that has resulted in the discovery of oil and its production from crystalline basement rocks in scores of hydrocarbon fields in the Caspian Sea region, western Siberia and the Dnieper-Donets Basin.

ABIOGENIC THEORY

Russian and Ukrainian scientists have argued that the standard account, calling for the formation of highly reduced high energy density hydrocarbons from highly oxidized low energy density organic molecules, violates the second law of thermodynamics and that the formation of such highly reduced molecules requires high pressures that are encountered only in the Earth's mantle. In 2002 Jason F. Kenney of the Gas Resources Corporation in Houston – a leading American advocate of this theory who also worked at the Russian Academy of Sciences – and his Russian colleagues published the result of experiments that produced an entire suite of petroleum fluids in a special apparatus that mimicked the conditions 100km below the Earth's surface by operating under high pressure (fifty MPa) and temperature (up to 1500°C). Although the paper appeared in the *Proceedings of the National Academy of Sciences* its publication did not lead to any widespread adoption of the theory.

Evidence for the biogenic origins of oil, an explanation shared by nearly all European and North American petroleum geologists, has been strengthened by the use of the latest analytical methods. Biomarkers in crude oil (including porphyrins and lipids) are clearly derived from organic molecules; isotopic analyses confirm matches

> ### ABIOGENIC THEORY (cont.)
>
> between carbon isotope ratios in hydrocarbons and in terrestrial and marine plants. Isotopic analysis of carbon and hydrogen rule out the occurrence of globally significant abiogenic alkanes. None of this means that there are no hydrocarbons of abiogenic origin or that we can satisfactorily explain the formation of all major oil and gas deposits. And it is always good to keep in mind that scientific paradigms shift, and that geology provides one of the best examples of this development: consensus opinion of geologists was to keep dismissing the existence of vigorous plate tectonics until that idea (championed by Alfred Wegener, a German meteorologist, since the 1920s) suddenly became a key paradigm of modern geology during the 1960s.

Geology of hydrocarbon deposits

Commercial viability of oil deposits is determined by the great trinity of hydrocarbon geology, the right combination of a rich source rock, a permeable and porous reservoir rock and a suitable tight trap to hold the liquid in place. The genesis of a source rock and sediment with a relatively high content of organic matter is controlled by interactions of three rates, those of production, destruction and dilution. All of them are, in turn, governed (often in a non-linear fashion) by a multitude of external factors. Each one can fluctuate across at least an order of magnitude, and only their appropriate combination will result in high accumulation rates. Additional factors that have a major influence on the rate of organic matter accumulation are the sulfurization of organic molecules and their adsorption onto (or within) clay particles: both processes make the molecules less vulnerable to oxidation and increase the probability of their preservation.

Primary (photosynthetic) productivity in marine environments depends critically on the availability of nutrients (in all aquatic environments particularly on the fluxes of nitrogen, phosphorus and iron) in the photic zone (part of the water column penetrated by sunlight) and on the concentrations of atmospheric CO_2. In turn, nutrient concentrations are determined by the intensity and extent of oceanic upwelling (transfer of deeper, nutrient-rich waters to the surface) and on nutrient transport by rivers, while the secular fluctuations of CO_2 (driven by orbital, geotectonic and climatic factors) often result in an obvious cyclicity of deposition.

High productivity is more likely to overwhelm the destruction rate and result in net deposition of dead biomass, but it alone could not guarantee substantial deposition as it could be largely negated by excessive dilution. Destruction of deposited organic matter (largely through its degradation by microorganisms and return of carbon to the atmosphere as CO_2) is primarily the function of oxygen concentrations: in oxygen-rich settings its rates are one or even two magnitudes faster than in oxygen-poor or anoxic environments. Consequently, the presence of anoxic bottom water has often been considered as the main controlling factor for organic matter preservation. The decay function is non-linear and only the settings with less than one milliliter of oxygen per liter had a chance to accumulate significant amounts of dead biomass.

Dilution, that is, organic carbon-free sedimentation, can be extremely variable, spanning four orders of magnitude and amounting to as much as $20kg/m^2$. Sedimentation that can create optimal conditions for the preservation and subsequent processing of accumulated biomass is a matter of a delicate balance between desirable rates of burying the organic matter (and hence isolating it from oxygen) and lowering the sediment's organic content below the levels capable of producing oil-rich source rocks. In marine ecosystems, high rates of dilution can also result from high photosynthetic productivity that contributes a relatively large mass of hydrogen-poor tests, shells and bones. For example, most of the mass of coccolithophorids (single-celled

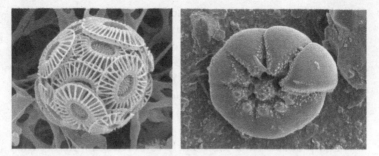

Figure 9 The intricate carbonate armor of a coccolithophorid (*Emiliania huxleyi*) and a foraminiferan protist (*Ammonia parkinsoniana*)

planktonic protists) is their intricately built calcium carbonate armor (see figure 9).

Accumulations of organic matter, and hence the generation of source rocks, were not spread evenly through geological eras but were clearly concentrated in a limited number of intervals whose duration and recurrence were determined primarily by global geotectonic cycles and secondarily by planetary climate changes driven by orbital oscillations. The total mass of organic matter stored in the Earth's crust is in the order of 10^{16} tons, with 10^{14}t (100 trillion tonnes) stored in organic-rich rocks, principally shales, containing at least 3% of organic matter (the range is from fractions of a percent to as much as 40% for oil source rocks and nearly 100% in some types of coal).

An approximate secular division of this global accumulation indicates that different geological eras contributed the following shares of the world's kerogens:

- Nearly 30% from the mid-Cretaceous period (about 100 million years ago)
- 25% from the late Jurassic period (around 150 million years ago)

- Less than 10% each from the late Devonian (350 million years ago), Silurian (408–438 million years ago) and early Cambrian (about 550 million years ago) periods

The mid and late Mesozoic eras are thus the source of more than half of all kerogen and hence, not surprisingly, of a similar share of oil originally in place in the uppermost sliver of the Earth's crust. Combustion of refined oil products is thus primarily converting a planetary patrimony that remained undisturbed for more than 100 million years.

The next step in the genesis of exploitable oil resources is the release of hydrocarbons from the kerogen beds where they were formed and the movement of crude oils (migration) from fairly (or highly) impermeable source rocks to relatively highly permeable reservoir rocks, and eventually to the Earth's surface and hence to oil's inevitable bacterially mediated decomposition. We will never know how much oil made it that far during the elapsed geological eras and was rather rapidly decomposed: we deal only with that still imperfectly known fraction that was confined in reservoir rocks by a variety of traps and that we discover by geophysical exploration and drilling (see chapter 3).

During the primary migration, oil is expelled from its source rock and, although water is always present, its transport through pores and capillaries does not always take the form of a solution or emulsion (oil being fairly immiscible) but it is one of an independent movement driven by the pressure of overlying rocks and made easier by the presence of faults and fractures. Oil expelled from its source rocks represents only a tiny fraction of migrating liquids (in the order of 10^{-2}%) and it is also highly dispersed, as organic-rich source rocks of a single formation may cover hundreds of square kilometers. Secondary migration, proceeding through a much more permeable media of porous rocks and driven largely by oil's buoyancy, concentrates oil by carrying it to one of the many possible traps where it accumulates.

Tertiary migration (leaks, seepages) leads oil from traps to the surface. When migrating, hydrocarbons lose their volatile components; they become heavy and non-liquid and the crust contains large volumes of these semi-solid or solid materials.

Virtually all commercially viable oil reservoirs are products of migration: *in situ* origin of oil in trapped reservoir rocks is highly unlikely as it would require the source rocks to be situated within traps. The word reservoir invokes images of large liquid pools but underground pools of oil are extremely rare. Instead, a reservoir is any subsurface body of rock whose porosity and permeability are sufficient to store and to transmit fluids so they are eventually able to flow into a borehole. Reservoir shapes are determined by the subsurface structures or discontinuities that envelop them or delimit them and they can range from relatively shallow, irregularly shaped near-horizontal lenses to steep, wedge-like enclosures.

Porosity is the share of void space in the reservoir rock that can be filled by hydrocarbons and, of course, also by water and non-hydrocarbon gases. High porosity may be created during the rock's deposition (primary porosity) or it can arise through a later alteration of the formation (secondary porosity due to recrystallization or fracturing). However, isolated pores would preclude oil extraction and hence it is the effective porosity, the volume of interconnected pores in a rock, that determines eventual productivity. Sedimentary rocks are generally much more porous than even the most fractured igneous or metamorphic formations, and sandstones and carbonates (limestones and dolostones) are by far the most common reservoir rocks. Their porosity ranges from less than 10% to about 70% (in some limestones), but typically 20–30%, of the rock's total volume.

Permeability quantifies a rock's capacity to transmit fluids. Permeable rocks have a multitude of relatively large and abundantly connected pores, but impermeable formations (shales being a good example) have much smaller and poorly connected pores. As with the porosity, effective permeability – the capacity to transmit a fluid when other fluids are present – is a more telling indicator. Permeability, measured in darcy units (D), can range over four orders of magnitude, from 0.1mD to more than 10D. Such impermeable rocks as salt, shales and anhydrite (calcium sulfate, a soft rock created by the evaporation of seawater) are able to keep reservoir rocks sealed through geological eras because their permeability is merely 10^{-6} to 10^{-8}D. Reservoir rocks with good porosity and permeability are classed in two broad categories, those of clastic and carbonate systems.

CLASTIC AND CARBONATE SEDIMENTS

Clastic sediments are formed from fragments of various rocks that were transported and redeposited to create new formations: sandstones, siltstones and shales are the most common examples. Clastic deposition leads eventually to sandstone reservoirs and many distinct processes result in such accretions. Alluvial fans accumulate different sizes of eroded matter as mountain streams spill onto open, flatter ground. This intermittent deposition has only rarely produced giant reservoirs: Venezuela's Quiriquire is perhaps the most notable example. Rivers deposit substantial volumes of sand and gravel particularly when they flow as braided or meandering streams. Alaska's Prudhoe Bay originated as a Triassic sandstone deposit in a braided flow.

Sedimentation in lakes has produced only a limited number of major oil reservoirs, most notably in China. In contrast, sedimentary rocks that originated as submarine fans harbor such rich oil reservoirs as the Forties field in the North Sea. But no other clastic depositional systems have contributed as many reservoir rocks as river deltas, ranging from elongated formations to wave-dominated forms: at least two-fifths of global oil production has come from deltaic sediments. Notably, the Tertiary deposits along the US Gulf coast, a number of Venezuelan coastal fields, the Niger delta and the south Caspian Sea. Shallow marine flats are also ideal depositional settings, often producing source rocks in association with reservoir rocks: fields in the central part of the North Sea have this origin.

Many of the world's largest reservoirs are in carbonate rocks. These are formed either inorganically, through a chemical reaction

CLASTIC AND CARBONATE SEDIMENTS (*cont.*)

(precipitation) of seawater calcium and carbonate ions in shallow seas, or by the process of biomineralization whereby marine organisms produce large carbonate structures, such as reef-building corals, or a constant rain of tiny shells – algal coccolithophorids and foraminiferal protists are the most common biomineralizers (see figure 9). Calcite and aragonite, two chemically identical compounds (calcium carbonate) with different crystal structures, dominate limestone composition. Partial replacement of calcium by magnesium (from evaporating seawater or during a deep burial) creates dolomite, the principal constituent of dolostones, whose higher porosity makes for excellent reservoir rocks. Carbonates can also be fragmented and redeposited as clastic sediments.

Most of these sediments originated in shallow seawaters (shelves) and their simplest subdivision is into ramps (with gently sloping sides) and platforms (flat, with sharply sloping sides). The world's largest oil reservoirs in the Persian Gulf were produced from Jurassic or Cretaceous shelf limestones. The Permian grainstones of West Texas and south-eastern New Mexico form the most productive onshore oil province in the continental US, while Alberta's giant oilfields are outstanding examples of production from Devonian reefs. Dolomitic reservoirs are also widely distributed, with the most productive ones (from the Upper Jurassic) in the Arabian platform. Deepwater carbonates are much less common (Mexico's Poza Rica and the North Sea's Ekofisk are outstanding examples), as are karst reservoirs (notably in buried hills in China).

Oil is contained in reservoirs by various traps, with rocks in the shape of domes or in angled formations that can originate through a variety of structural and sedimentary processes and that are sealed by highly impermeable rocks, most often by shales or evaporites (see figure 10). Structural traps are formed by deformations of the Earth's crust. Anticlines, smooth arch-shaped (convex) folds, make such outstanding hydrocarbon traps that they enclose nearly four-fifths of the world's largest oil reservoirs. Some anticlinal traps have reservoir rocks situated right in the middle (the core) of the convex shape where they are capped by

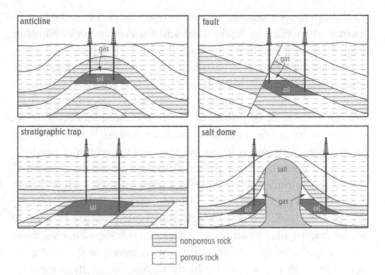

Figure below the images shows labels: anticline, fault, stratigraphic trap, salt dome with gas/oil markings, and a legend:

- nonporous rock
- porous rock

Figure 10 Four common kinds of oil traps

impermeable strata. Anticlinal structures can be both large and almost perfectly symmetrical (seen as ovals on topographic maps) or can be closed by faults (relative displacements of rock strata that prevent further oil migration) including normal, reverse and strike-slip arrangements.

The most extensive structural traps are formed by compressive tectonics along convergent plate boundaries. Such contractional folds, created during the tectonic collision of Arabia with Eurasia, produced an entire system of two spectacular asymmetric whale-back anticlines (Lurestan and Fars) in the Zagros Fold-Thrust Belt of Iran. In between is the Dezful Embayment – a depression some 600km long, running in a NW-SE direction, and up to 200km wide with a total area of some 50,000km^2 – that harbors forty-five oilfields, including such supergiant reservoirs as Agha Jari, Ahwaz, Bibi Hakimeh, Gachsaran and Marun. Source rocks of these fields are an early Tertiary Pabdeh formation and a mid-Cretaceous

Kazhdumi deposit, and the two main reservoir rocks are the Sarvak formation (overlaying Kazhdumi) and the Asmari rocks. Another large-scale anticlinal trap contains oil in eastern Venezuela.

Other remarkable structural traps are created, albeit on a smaller scale than tectonic anticlines, by diapirs, vertical intrusions of lighter formations through denser rocks. Because salt (halite) has a lower density than the surrounding sedimentary rocks (2.2 vs. at least 2.5g/cm^3) it is buoyant and as it rises through sedimentary formations it can become deformed under pressure, creating domes, sheets or pillars. Anticlinal structures generated by rising salt domes create excellent traps on their own and they are also often associated with evaporite rocks (gypsum, anhydrite) that provide perfect caps and seals. Reservoir rocks are frequently found draping over a rising salt, with steep side dips. Diapiric anticlines can also be formed by rising hot magma (but the heat generated by their rise would be likely to destroy the hydrocarbons present in the reservoir rock), shales and movements of mud. The world's second largest oilfield, Kuwaiti al-Burqān, is a Cretaceous sandstone trapped above a massive (roughly 750km^2) salt swell.

Stratigraphic traps are generally smaller, resulting from gradual accretions of impermeable rocks that enclose oil-bearing formations (depositional traps) and from sudden unconformities (created by erosion or by karst phenomena). They also arise from mineral precipitation, dissolution, intrusion of tar mats or permafrost. Primary stratigraphic traps include such elongated sandstone bodies as point bars, deltaic channels and barriers as well as carbonate slopes and coral reefs. Secondary stratigraphic traps come as various unconformities, clay-filled channels of diagenetic dolomites or calcites. Stratigraphic traps are often present as multiple, complex, staggered phenomena, and they are also often combined with structural traps. Major oilfields with stratigraphic traps include Alaska's Prudhoe Bay, East Texas, and Bolívar Coastal, Venezuela's supergiant along the east coast of Lake Maracaibo that is bounded by an enormous tar seal.

AL-GHAWĀR OILFIELD

Some of the principal concepts of petroleum geology that have been introduced in this chapter are illustrated here in describing the essential features of al-Ghawār, the world's largest super-giant oilfield that was discovered in north-eastern Saudi Arabia about 80km inland from the western coast of the Persian Gulf. Al-Ghawār's extraordinary size and its high sustained productivity are due to a nearly perfect combination of a large structure sourced from a prolific formation into a highly permeable reservoir capped by an excellent seal. The field is a large (250km long and about 30km wide) gentle anticline with two subparallel crests (separated by a saddle); the anticline is folded over, an elevated deformation of the basement rock that arose originally during the Carboniferous period (see figure 11).

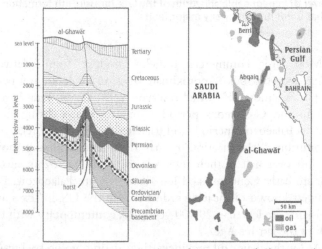

Figure 11 East-west cross-section of strata at al-Ghawār, the world's largest oilfield, and the field's extent

Both the source and reservoir rocks belong to the Upper Jurassic Shaqra group. Al-Ghawār's source rock is a highly prolific, organic-rich lime mudstone of the Tuwaiq Mountain Formation (with about 3.5% of total organic carbon), about 150m thick, that

AL-GHAWAR OILFIELD (cont.)

was laid down in intershelf basins between 164 and 155 million years ago. The overlying Hanifa formation (with about 2.5% of organic carbon) also contributed to the original store of organic matter. Because of the great oil volume in place it is likely that a major lateral migration took place before the oil migrated a short vertical distance to the reservoir rock, late Jurassic Arab-D limestone laid down between 150 and 145 million years ago that lies 1.8–2.1km below the surface.

This rock is up to 120m thick and its quality improves from the bottom mudstone to clean top grainstone. Its porosity exceeds 30% in some parts of the reservoir (mean of 14–18%) and its permeability is also excellent (in parts more than 600mD). The oil-bearing limestones are capped by a massive (up to 150m thick) layer of impermeable anhydrite of the late Jurassic Hith formation that is not broken by any major faults.

Most large commercial oilfields produce from relatively young reservoir rocks. Approximate division by geological periods places almost 45% of all reservoir rocks in the Jurassic and 35% in the Cretaceous period while reservoirs dating from the late Palaeozoic periods (Silurian to Permian) are relatively uncommon. A classification by major producing oilfields shows that just over half of their reservoir rocks are Mesozoic, nearly 40% are early Cenozoic and less than 10% are Palaeozoic. The youngest known oil-bearing sediments (in the US, in Texas and Louisiana) are less than 20,000 years old, contemporary with the peak of the last Ice Age.

As for the major oil provinces, their dating (going backwards in time) is as follows:

• Mid-late Cretaceous source rocks are found in China's Songliao Basin (the site of Daqing, the country's largest supergiant field), in the Gulf of Mexico, the Persian Gulf, Venezuela and in waters off Congo

- Upper Jurassic layers are productive in Western Siberia, the Persian Gulf and the North Slope of Alaska
- Late Jurassic rocks occur in the Gulf of Mexico, the North Slope of Alaska, the North Sea, Western Siberia, the Persian Gulf and Northern Caucasus
- Triassic, early Permian to late Carboniferous basins include China's Tarim and Junggar and the southern part of the North Sea
- Early Carboniferous to late Devonian sediments produced such important oil provinces as Russia's Volga-Ural and Timan-Pechora, the West Canadian Sedimentary Basin, Oklahoma and Texas

This distribution, with commercially exploitable resources of oil present on all continents, makes it difficult to make any grand generalizations. One intriguing observation is that the world's largest oil-bearing basins are concentrated either along the shores of the Mesozoic Tethys Ocean that was wedged between the supercontinents of Laurasia (to the north) and Gondwana, extending from today's Indonesia through the Middle East to North Africa, or along both flanks of the American Cordillera, from Alberta through the Gulf of Mexico to Trinidad, Venezuela and Brazil and, on the western side, from Alaska through California to Ecuador. On the other hand, size distribution of the world's oil resources makes it possible to make a number of fascinating observations.

The size of oil reservoirs shares a highly skewed (geometric) frequency distribution with many of the Earth's discrete physical features: for example, there is only one Greenland, and just two islands of the size of New Guinea and Borneo, thousands of small islands and myriads of tiny reefs and islets. Similarly skewed frequency distributions can be plotted for the size of lakes or the length of peninsulas – or for the sizes of oilfields: most of the world's commercially exploited oilfields hold only small volumes of oil (and there is, obviously, a much larger number of even smaller reservoirs that are not worth exploiting) while a small number of giant and supergiant reservoirs contain a disproportionately large share of the oil present in the Earth's crust.

By the beginning of the twenty-first century there were – depending on how some contiguous or adjacent structures are classified, and also on the uncertain estimates of ultimately recoverable oil – at least 500 giant oilfields (containing more than 0.5Gb of ultimately recoverable resources) in twenty different regions. Between 2000 and 2014 new discoveries added 90 giant oilfields and 12 combined oil and gas giants. Giant oilfields contain at least two-thirds of all conventional crude oil reserves, and the supergiants alone (each with more than 5Gb) claim nearly one half. Five of the largest conventional supergiants – Saudi al-Ghawār, Kuwaiti al-Burqān, Mexico's Cantarell, Venezuela's Bolívar and the Saudi Safāniya-Khafjī – have approximately a quarter of the world's oil reserves. I must stress that all of these shares are not only rounded approximations but that they also change with continuing reappraisals of ultimately recoverable oil in producing fields; these are the consequence of additional drilling and higher recovery rates and commonly result in substantial additions to initially estimated reserve totals.

Spatial distribution of the world's largest oilfields is also highly skewed: the Persian Gulf region contains more than half of all supergiants and nearly 30% of all giants. Western Siberia has about 12% and the Gulf of Mexico nearly 10% of all giants. The regions with approximately 5% of the world's giants include the Anadarko/Permian Basin in Oklahoma, Texas and New Mexico, the Volga-Urals, the Caspian Sea, and South-East Asia (Thailand, Indonesia and Vietnam). Regions with fewer than ten giants include North Alaska, Brazil, the Black Sea, Siberia, north-western Australia and the Bass Strait. Classification based on the basin type shows that about 30% of giants are situated along passive continental margins, another 30% are along continental rifts and nearly 25% are found along continental collisional margins. In the future, there may be some pleasant surprises as we discover more giants (the most promising regions are the Bay of Bengal, China's Ordos and Tarim basins, the Mekong Delta, Sudan and waters offshore from northern Australia) but these discoveries will not radically change the global pattern of oil distribution as the Middle East will retain its dominance.

3

How oil is found and where it has been discovered

The search for mineral resources is an inherently difficult and commercially risky enterprise. Compared to many deposits of metallic ores that are found in thin and twisting veins deep underground, many oil reservoirs are massive structures that lie fairly close to the surface. Some, signaled by surface pools, seeps or gas vents, were discovered without any of the sophisticated geophysical tools and procedures that are available to modern explorers, but many remarkable innovations were required to find smaller and deeper reservoirs. These advances have ranged from better theoretical understanding of the geology of oil-bearing formations and clever exploratory techniques to complex three- and four-dimensional (3D, 4D) simulations and visualizations of oil reservoir dynamics.

Eventually, exploration wells have to be sunk to verify the presence of significant volumes of oil, but given the cost of drilling (increasingly in remote continental locations that are often very difficult to access, in deeper and stormy offshore waters, and at greater depths, both onshore and in what is now called ultradeep ocean) it has become imperative to minimize the risk of repeatedly ending up with dry wells. But even with the advanced geophysical methods that provide unprecedented knowledge about the formation to be drilled, every new exploratory well (and particularly those in promising but previously undrilled locations) still justifies the oil industry's traditional name for such an undertaking, a wildcat. The main reason for this is that none of the parameters of an explored oil formation, and particularly not its pressure configuration, are known with certainty and drilling crews must be prepared for unexpected kicks (inflows of reservoir fluids, water, oil or gas, into a well bore during its progress) that may lead to violent well blowouts.

After a brief history of oil exploration and the drilling methods used to discover oil, and a more detailed description of current drilling tools and practices, I will turn to historical accounts of oilfield discoveries. Most of the drilling – starting with the world's first wells that were completed explicitly in search of oil during the 1850s, proceeding through the pioneering exploration decades of the late nineteenth century, culminating in the discovery of most of the giant oilfields that are in production today, and continuing in the still inadequately explored regions of the Earth – has been done on land, but many remarkable advances in exploratory offshore drilling created a new (post-WWII) industry without which the world would have missed nearly 30% of its recent crude oil supply.

Geophysical exploration

Early explorers for oil had little to guide them: obvious surface signs of oil's presence (such as natural seeps or pools, tarry rocks or tar lakes) were uncommon, subtler geochemical indicators were not well understood, and hunches and guesses were a costly way to decide where to drill. But eventually oil exploration adopted and adapted techniques of geophysical investigation, above all the studies of the Earth's natural electrical, gravitational and magnetic fields and of the propagation of seismic (elastic) waves through the Earth's crust. Reflection seismology, a key exploratory tool, maps subsurface deposits by clocking the time needed for a pulse sent underground to return to the surface after it has been reflected from the interfaces formed by different types of rock formations.

HISTORY OF SEISMIC EXPLORATION

The seismic search for oil had its origins in Reginald Fessenden's research on detecting icebergs and locating ore bodies (patented in 1917) and in the independent invention of practical, albeit initially crude, seismographs. The first portable device was invented by Ludger Mintrop during World War I in order to give the German army a highly accurate means of locating the positions of Allied artillery. In the US John C. Karcher developed a reflection seismograph at the federal Bureau of Standards after the end of World War I and the first field test of the device (on June 4, 1921) confirmed that seismic techniques can unveil the existence of oil-bearing structures. In 1926 his crew located a promising structure around Seminole in Oklahoma, and on December 4, 1928 the first well ever to be drilled by following the results of reflection seismography struck oil. In 1930 Karcher founded the Geophysical Service Company which was to become Texas Instruments (later, a leader in microelectronics).

Reflection seismography became much more sensitive with the use of vacuum tubes and, since the early 1950s, with the adoption of transistors. The decade's other innovations included electronic data recording and processing, Harry Mayne's signal-to-noise enhancing technique, and Conoco's vibroseis method that substituted waves created by vibration or weight dropping for waves generated by dynamite explosions. In the late 1960s, Mike Forrest, a Shell geophysicist, discovered that strong seismic reflections ('bright spots') on the crest of hydrocarbon-bearing formations make it possible to use seismic data for the direct detection of oil and gas formations. Digitalization and computerization of seismography also started during the 1960s and advanced during the 1970s; at the same time, oil exploration's enormous demand for data processing was among a few key factors that pushed the performance of new hardware and required complex new software. The most remarkable sign of this interplay was the introduction of 3D seismic surveys. The acquisition and technique was pioneered by Exxon in the Friendswood oilfield near Houston in 1967 and it became routine by the mid-1970s.

All seismic surveys use the same basic procedure: they generate sound waves at the surface (using truck-mounted vibration pads or, where trucks cannot be driven, dynamite charges in shallow holes) and an array of sensitive receivers (geophones) to record the sound waves reflected from rock formations. In marine exploration, the sound waves are generated from air guns (firing compressed air) that are towed on a line behind a ship and reflections are recorded by hydrophones located further down the line (see figure 12).

Standard three-component seismic surveys determine the type of wave and the direction of its propagation by using three orthogonally oriented geophones; hydrophones are added to ocean-bottom sensors to measure additional waves for a four-component seismic survey. The massive amounts of data are then processed by high-speed computers. The resulting 3D maps provide unprecedented insights into the structure of subsurface formations and hence inform on the best ways to exploit them. Increased computing power (driven by Moore's famous law of doubling the capacity of the microprocessor every two years) was an essential ingredient of the 3D seismic revolution as the amount of raw data that had to be interpreted for a typical survey increased more than 5,000-fold!

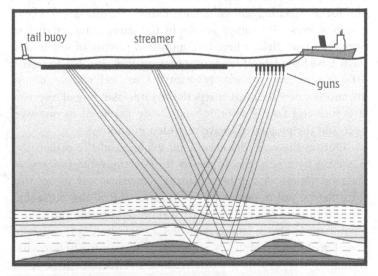

Figure 12 Simplified diagram of a marine seismic survey

Another computing advance that revolutionized oil exploration and called for an unprecedented amount of data processing capacity was the development of 3D visualization. Arco, Texaco and Norsk Hydro were the first companies to use large immersive visualizations whereby the experts are enveloped by tall projection screens or, even more impressively, enter visualization rooms where they are literally surrounded by data through which they can walk and while virtually immersed in an oil reservoir they can choose the best location and paths for new wells. And this process has been carried yet another step forward with the development of time-lapse 4D seismic surveys.

Repeated monitoring of a reservoir over a period of weeks or months unveils its changing properties, above all its temperature and pressure and the paths the fluids (oil, water, gas) take as they move through it. This knowledge is invaluable for anticipating future reservoir flows and hence calculating the site of new production wells, for identifying oil- or gas-bearing areas that have been bypassed by existing wells and that are worth returning to, and for accelerating and enhancing existing hydrocarbon recovery. Not every 4D survey results in the gains demonstrated in Sumatra's Duri field, where the optimized placing of steamflood wells raised the rate of reservoir recovery from just 8% to nearly 60%, but rewards are always substantial. Crosswell seismic surveys are another new tool that sharpens the understanding of reservoir structure and flow – acoustic sources are deployed in one well bore and the propagated wave recorded in another.

During the early decades of oil exploration the drillers did not even have any reliable means to determine the progress of their operations. A cumbersome and expensive way to get that information was to periodically use a special drill that cut a slim cylindrical core of solid rock that was brought to the surface for examination. This is still done when a periodic examination of actual rock core samples is required: an expensive hollow diamond-covered drill is used to remove cylinders of rock for a preliminary appraisal in the field or more detailed evaluation in a laboratory. A cheaper alternative is sidewall coring, using either a small explosive charge to propel a tiny core barrel sideways from the well or a small robotic core bit to drill short (up to 5cm), thin rock core samples withdrawn by a wireline.

Recording the kinds of rocks and their attributes based on the samples of rock cuttings (so called mud logs) is an easier option than coring, and this logging, together with the measuring of the rate of penetration, became the first step toward getting a systematic record of the changing properties of rock formations. Eventually these records became automated and modern well logs inform on variables ranging from the speed of rotation and hoisting to the mud-flow rate and pressure and temperature at the bottom of the well. The single biggest advance in oil exploration came in 1911 when Conrad Schlumberger, at that time a lecturer at the French École des Mines, came up with the novel idea of using electrical conductivity measurements in prospecting for metallic ore deposits. A year later his first crude map of equipotential curves indicated that the technique could also be used to identify subsurface structures that could act as traps for such mobile resources as oil and gas.

The process of electric well logging pioneered by Schlumberger is one of the mainstays of modern geophysical exploration: it consists of measuring a number of revealing physical variables in or around a well and recording them for subsequent evaluation.

WELL LOGGING

The *Société de Prospection Electrique*, the precursor of today's Schlumberger Company, was set up in 1926 and it rapidly expanded its operations to the Americas, Asia and the USSR. In 1927 Henri Doll, an experimental physicist, produced the first electrical resistivity well log, a record of successive resistivity readings that are

WELL LOGGING (cont.)

used to create a resistivity curve. In 1930 the company introduced a continuous hand recorder, and a year later Conrad, together with his brother Marcel Schlumberger and E. G. Leonardon, recognized the phenomenon of spontaneous potential that is generated naturally between the borehole fluid electrode and the formation water in permeable beds. Simultaneous recordings of spontaneous potential and of the resistivity curve could differentiate impermeable and permeable strata, and hence identify potentially oil-bearing beds.

In 1949 Henri Doll introduced the first induction logging, that is, the measuring of electrical resistivity in a borehole after inducing alternating current loops in a formation that made it easier to distinguish oil-bearing rocks from water-bearing substrates. Porous formations that contain only salt water have very low resistivity, often less than 10 ohm-meters (Ωm) and those that might harbor hydrocarbons have high readings (>50Ωm). Well logging became an indispensable tool for appraising the probabilities of oil discovery, and Schlumberger has retained global leadership in the field by continuing to introduce improved devices and acquiring allied companies. Modern logging uses slim, long (6–25m) pipes to house the instruments; these tools are lowered into a new borehole on a flexible electric cable (wireline) spooled out from a special truck and transmit the readings to the surface.

Gamma-ray logs record the natural emissions from radioisotopes present in the formation, thereby distinguishing between shales and sandstones that have different gamma-ray signatures. Monitoring of the formation's density is based on the decline of gamma-ray flux between a source of radiation and detector; in porous rocks more of the emitted rays reach the counter than in dense formations. Neutron logging is used to evaluate formation porosity by measuring its effect on fast neutrons emitted by a source. Logging is normally done after a well is drilled but logging-while-drilling is now also possible thanks to making various sensors (monitoring well inclination, pressure, resistivity, density and porosity) part of an integrated bottomhole assembly, the lower portion of the drill string joining the drilling bit to the drill collar, heavy drillpipe and crossover pipes. Logging-while-drilling is particularly useful where using the standard wireline tools is difficult or outright impossible (particularly in highly deviated wells).

Despite impressive advances in geophysics, seismic surveys and computerized data processing and imaging, modern oil exploration cannot eliminate the risk of drilling dry holes. And this risk is uncommonly expensive: with deeper drilling on land and deep offshore wells, a dry hole onshore can cost 'just' $6–10 million (but it could be easily twice as expensive), and while leasing an offshore drilling rig cost around $150,000/day in 2006, the averages in 2016 were between $350,000 and $400,000 a day. Unlike in many other industrial endeavors, nothing can be salvaged from a miss, although the cost of the failed enterprise can be written off as a business expense (in the US, in full in the same year). At the same time, oil exploration remains (despite the reduced probabilities of discovering supergiant or giant fields) a highly rewarding enterprise. Even with a relatively low price of $50/b the discovery of the smallest giant oilfield (containing half a billion barrels of recoverable crude) represents a find worth $25 billion just in terms of crude oil value, and a smaller field with an ultimately recoverable volume of 20Mb would be worth at least one billion dollars.

The world's most intensively explored oil provinces included all basins in 48 coterminous US states as well as Alaska, Alberta and Saskatchewan, the North Sea and European Russia. Coastal Venezuela, south-western Iran and the seas off north-western Australia belong to oil basins explored with moderate intensity. Partially explored regions include the shelves of West Africa, parts of the Mediterranean and some interior parts of China. Relatively unexplored regions with considerable oil production potential include parts of eastern Siberia, large areas off the east coast of Latin America, West Africa and Greenland, and large expanses of Arctic and Antarctic waters (and, of course, Antarctica itself, but the 1991 international treaty forbids any mining activities).

Not surprisingly, exploratory drilling activity is greatly stimulated by higher oil prices, while their precipitous retreat can lead to drastic retrenchments. And there is always the risk that by the time someone makes a well-considered decision to drill (after becoming convinced that high oil prices are more than an ephemeral phenomenon) an unforeseeable event, or a foreseeable reversal taking place much sooner than anticipated, can turn a cautious and highly promising investment into an instant and completely irretrievable loss. US statistics illustrate these price-driven fluctuations. In the early 1970s, before OPEC's rapid price rise in 1974, slightly more than 1,000 rotary rigs were searching for oil and gas on land and offshore; the subsequent price rise pushed that total to more than 4,500 by the end of 1981, but by 1985 the count was below 2,500 and by the middle of 1986 (in the wake of the price collapse) it was below 700. Subsequently the total stayed mostly below, or just above 1,000 until a slight rebound that began in 2004. The number of rigs briefly surpassed 2,000 in September 2007, and again in late 2011 and early 2012, before sinking to a new low of just 408 wells in May 2016. In 2014 about 104,000 new wells were drilled worldwide onshore (45% of them in the US) and almost 3,400 offshore (less than 10% of them in the US), while a year later, as oil prices retreated, the onshore drilling was down by about 7% and offshore exploration declined by about 9%. The US remains the new exploration leader with China a distant second, followed by Canada and Russia.

Oil drilling

Regardless of whether the drilling was done in order to explore for oil or to develop newly discovered oilfields and then to maintain and enhance their production, there have been only two dominant techniques used to complete the vast majority of wells since the beginning of the oil era during the 1850s: percussion (cable-tool) drilling during the first half of this period, and rotary drilling afterwards. Remarkably, the technique that was used to sink the first wells explicitly drilled in search of oil has its origin in an ancient Chinese invention. Percussion drilling was used from at least the beginning of the Han dynasty (Rome's great contemporary, constituted about 200 BCE) to extract natural gas in Sichuan province (see figure 13).

The gas was then transported by bamboo pipes and burned under large cast-iron pans to evaporate brines and produce salt, a precious commodity in the landlocked province. The method required a congruence of several important technical skills: availability of heavy iron bits and long bamboo poles, manufacturing of strong and long cables woven from bamboo fiber, and a clever use of levers. Heavy iron bits were attached to long bamboo cables that were suspended from bamboo derricks and they were then repeatedly (and rhythmically) raised and dropped into

Figure 13 Qing dynasty drawing of the ancient Chinese technique of percussion drilling using a bamboo derrick

a manually dug hole by two to six men jumping on a lever. This seemingly primitive way of drilling was actually quite effective and its performance improved with use. Han dynasty boreholes made by percussion drilling were only about 10m deep but by the tenth century wells deeper than 100m were common and in 1835 the deepest well reached 1km below the surface! Until 1895 a modernized variant of this ancient Chinese technique was used to drill all new oil wells. Starting with Edwin Drake's first Pennsylvania well in 1859, small steam engines, rather than human muscles, powered the drilling process and heavy metal bits were hung from manila ropes and, later, from multiple-strand steel wires supported by small wooden derricks. Steam power and better bits speeded up the fracturing and pulverization of the drilled substrate but the bailing out of the cuttings accumulated at the bottom of the hole remained laborious. However, the technique remained in use for decades after the introduction of a superior drilling method: cable tool rigs outnumbered rotary drills until 1951.

The first rotary drilling rig was used in the Corsicana field in Texas in 1895, but in the US the technique became significant only with the completion of the record-flowing (up to 80,000bpd) Spindletop well in Texas in 1901, and it became dominant worldwide only after 1950. The fundamental importance of rotary drilling for both discovering oil and producing it from commercial wells justifies more than a cursory description of its main components and operating procedures. The entire assembly of structures and machines used in rotary drilling is called a rig and its hardware consists of six interconnected systems: power generation, hoisting mechanism, rotary table that turns the drilling pipes and the drilling bit, fluid circulation, blowout prevention and monitoring and well bore data acquisition.

Rigs are visible from afar thanks to their tall, pyramidal derricks, the steel structure that supports the crown block and the drill string (see figure 14). The crown block is a combination of fixed pulleys (sheaves) and travelling blocks used to gain a large

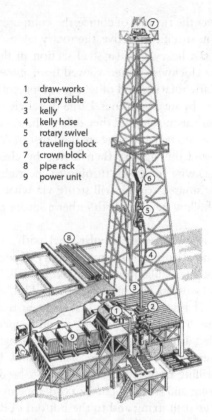

1 draw-works
2 rotary table
3 kelly
4 kelly hose
5 rotary swivel
6 traveling block
7 crown block
8 pipe rack
9 power unit

Figure 14 Modern rotary drilling rig (courtesy of BP)

mechanical advantage to hoist heavy loads with a small-diameter (1.9–3.75cm) steel cable; the draw-works reel the cable over the crown block and thus lower (by gravity) or raise (by electric motors or diesel engines) the travelling block and the load suspended from it, be it drill string, well casing or well liners. The drill string is made of the drill pipe and the attached bottomhole assembly that holds a drill bit. Other key parts of a rig are the tanks holding the drilling fluids and the pumps to dispense it, engines to power the rig and, of course, the component that gives the technique its specific adjective, the rotary table.

This table is a heavy circular steel section of the drill floor that powers the clockwise (when viewed from above) motion of a drill string. Early rotaries (and other rig equipment) were energized (via gears) by steam engines. Large diesel engines are now the dominant prime movers and they deliver their torque directly (via drive chains and belts) or they are used first to drive electricity generators and the rigs are then driven by electric motors. The table's clockwise rotation ('turning to the right' in drilling parlance) is transmitted to the drill string via what is known as a kelly, a long hollow steel bar with either a square or hexagonal profile.

Drill pipe is a section of tubular steel with threaded ends (tool joints) whose length and diameters vary depending on the requirements of a particular job; standard length is 9.44m (31 feet) with nominal diameters ranging from 7.3 to 11.43cm (that is, 2 and 3/8 to 4 and a half inches). Its first segment is attached to a kelly and as the drilling proceeds and a well deepens, new sections of threaded drill pipes are connected to a lengthening drill string which serves as a conduit for drilling fluid. To add these sections the drilling must stop, the kelly must be disconnected from the top joint and a new section of drill pipe must be mated to the top of the drill string and to the bottom of the kelly. The rate of penetration (typically in the order of 100m a day) will determine the number of new connections that can be made in an hour or a day.

The bottomhole assembly attached to the lowermost portion of the drill string consists of a bit, crossovers (sections of tubular steel used to join pipe with different thread or diameter) and heavy drill collars to keep the entire rotating assembly vertical. The cost of these assemblies ranges from about $150,000 for the simplest arrangement to more than ten times that for complex units that contain mud motors powered by drilling fluid and used in directional and horizontal drilling. When a drilling bit wears out, the entire drill string has to be withdrawn from the well, and pipe sections have to be unscrewed, stacked aside and then reattached after a new bit is mounted. This process (called tripping) is laborious and dangerous, as heavy pipes are removed, added and manhandled within the limited work area of a rig floor. The earliest rotary drills used fishtails and circular-toothed bits that were effective only in soft rocks. The invention of the rotary cone drill changed that dramatically.

Roller cone bits have either milled teeth or carbide tungsten inserts with embedded industrial diamonds (used in oil drilling for the first time in 1919) whose superior hardness gouges out the rock as the bit rotates. Diamonds are also used in drag bits that have no roller bearings and shear the rock with a continuous scraping motion. Polycrystalline diamond cutters use thin (about 3mm) circular bits (up to 2.5cm in diameter) of synthetic diamonds bonded to the underlying tungsten carbide and are preferred for drilling shale formations. Diamond matrix drag bits are used in extremely hard formations. Only about 20% of diamonds are sold to the jewellery trade; most of the rest go into drilling for hydrocarbons and metallic ores.

HISTORY OF ROTARY CONE DRILLS

In 1901 Howard Robard Hughes, who had failed to finish law school and had been working in ore mining, was so impressed by the discovery of the Spindletop oilfield that he switched to oil drilling in Texas. In 1907 he was unable to penetrate extremely hard rock in two promising localities and this failure made him determined to design a better drilling bit – and he did so over just two weeks in November 1908 while visiting his parents in Iowa. His truly revolutionary design was one of the most consequential patents of the twentieth century, granted on August 10, 1909. The rotary cone drill (see figure 15) consisted of two 'frusto-conical shaped rollers having longitudinally extending chisel teeth that

HISTORY OF ROTARY CONE DRILLS (*cont.*)

disintegrate or pulverize the material with which they come in contact and thus form a round hole in said material when the head of the drill revolves.'

The two rollers were arranged at an angle to each other as they rotated on stationary spindles and, of course, as the entire bit rotated at the end of the drill string. This simple, elegant and near-perfect design made it possible to drill ten times faster than with standard fishtail bits. In 1909 Hughes went into partnership with Walter B. Sharp to set up the Sharp-Hughes Tool Company (after 1918 just Hughes Tool) that made the bits, not for sale but for lease, at $30,000 per well. The company used its rising profits well as its engineers kept on improving the basic Hughes tool and developed new modifications. These included a design that was thought to be impossible: in 1933 Floyd L. Scott and Lewis E. Garfield patented a tricone bit, interfitting the cutters of three rotating drill cone bits (see figure 15). This arrangement sped up the drilling, offered a much better support on the well bottom than did the two-cone tool and reduced vibration. Numerous variants still dominate the oil drilling market and the Hughes Company (now Hughes Christensen, part of Baker Hughes) is still their leading producer, followed by Halliburton, National Oilwell Varco, Schlumberger and Varel International.

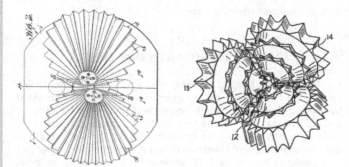

Figure 15 Drawing of the first conical drilling bit in Howard Hughes's 1908 patent application (left), and the first three-cone bit devised in 1933 by Floyd L. Scott and Lewis E. Garfield of the Hughes Tool Company

Besides the kind and the quality of the drilling bit, the other key variable that controls the rate of penetration through the drilled rock is the weight on the bit, that is, the mass of the drill string. Every combination of the bottomhole assembly and the drilled rock has the optimum weight on the bit that will produce the highest rate of penetration: this total is typically between 6.5 and 9t. Because the single standard drill pipe weighs about 190kg this optimum is achieved with only 34–47 sections, that is, with total drill string length as short as 300m. Here is the reason for the already described large mechanical advantage afforded by the wire led over the crown block at the top of the derrick and travelling blocks to the draw-works: this arrangement keeps the bit weight at the level commensurate with the fastest rate of penetration by suspending the remaining mass of the drill string. Depending on the number of loops between the two blocks the suspended load can be easily 500t, enough to support more than 2,000 standard pipe lengths that would reach more than 20km below the surface, much farther than even the deepest experimental well ever drilled (most oil wells do not go deeper than 4km).

Rock cuttings produced by an advancing drilling bit are removed from the borehole by a drilling fluid, commonly called drilling mud. This term is a collective label for what has become an enormous variety of liquids and mixtures of liquids, solids and gases. The simplest way to sort out their confusing variety is to classify them into water-based, oil-based, gaseous and synthetic fluids. Water-based muds are natural mixtures or fluids to which lignosulfonates, phosphates, lignite or tannins are added to act as deflocculants and viscosity and filtrate reducers; various emulsifiers, defoamers, polymers, salts, corrosion inhibitors and weighting materials (barite, iron oxide) are also used to control downhole pressure and prevent cave-ins. Oil-based muds are pumped into high-temperature wells or into deep holes where sticking and hole stabilization may be a problem. Synthetic fluids are primarily composed of esters, ethers and olefins. What started simply as the use of water to remove rock cuttings has evolved into a complex industry dealing with scores of varieties of drilling fluids, whose cost is typically 5–10% of the total cost of drilling a well.

Drilling fluid is pumped under high pressure from a storage container through a rigid metal standpipe that reaches about one-third of the way up the derrick into a kelly hose, a flexible high-pressure conduit (typically 7.5–12.5cm inside diameter) that is connected to the swivel. The swivel must be strong enough to suspend the weight of the entire drill string and it also allows the flow of pressurized drilling fluid from the flexible hose to a rotating kelly. Drilling mud is forced at high pressure down the kelly into the drill string and eventually it flows through the bit. This constant flow cools the rotating bit, making the removal of cuttings a continuous and easy operation, and the pressure it puts on the well sides helps to prevent the hole from caving in. The mud loaded with drilling debris returns to the surface through the annulus, the space between the outer drill string wall and the wall of the well bore.

Heavy drill collars are used to add weight and rigidity to the drill string in order to keep the borehole straight, and stabilizers may be used at intervals along the string to improve rigidity and to prevent deviation from the vertical. Wells will commonly encounter strata saturated with water, oil or gases or layers of unconsolidated sediments that must be isolated from the progressing borehole (and from each other) by the installation of steel casing that must be fastened in place. Cement was used to do this even in some pre-1900 wells drilled by cable tools, but the first machine to eliminate laborious hand mixing at the drill site, and the process that made the cementing of a well casing into a routine operation, came only in the 1920s.

As in the Schlumberger case, the company that pioneered and commercialized improved versions of this important

operation has retained its primacy ever since the 1920s. In 1919
Erle P. Halliburton started his oil-well cementing company in
Oklahoma and in 1922 he filed a patent for his new cement jet
mixer. Eventually Halliburton expanded into one of the world's
largest oilfield service companies (providing not only cementing
but also logging and well-completion services as well as drill bits)
and because of its merger with Dresser (another major energy
industry provider) and the acquisition of Brown & Root (build-
ers of the first offshore oil platform in 1947 and a large general
contractor) it is now also one of the world's leading multifunc-
tional engineering enterprises.

As the exploratory wells deepened and began encountering
higher reservoir pressures it became imperative to devise new
means of preventing well kicks and blowouts during drilling,
tripping or when the drill string is out of the well bore. A kick
takes place anytime the pressure in the well bore becomes lower
than the pressure of the surrounding formation fluids. Kicks are
caused either by insufficient mud weight, be it because of initial
low mud density or its lowering by an inflow of lighter liquids
or gases (an underbalanced kick), or because of sudden pressure
changes caused by the movement of the drill string or casing (an
induced kick).

Violent blowouts not only eject the drill pipe from the borehole but lead to fires that can kill crew members and destroy a rig.
The first blowout preventer, designed by James S. Abercrombie
and Harry S. Cameron in 1922, could contain formation pressures of up to 20MPa in 20cm diameter holes; in contrast, some
of today's high-performance preventers used to drill in geopres-
sured zones can contain formation pressures up to 100MPa in
boreholes with diameters up to 46cm. Blowout preventers must
be able to shut the well at the surface by using various ramming
devices, rid the well bore of the unwanted fluid and replace it
with a heavier fluid that will prevent future intrusions. Ever since

these devices have been widely deployed accidental blowouts have become increasingly rare. New measures and procedures were also put in place to minimize the damage of an inadvertent blowout and to allow for a safe and rapid evacuation of drilling rigs, precautions that are particularly critical in offshore oil drilling. But when accidents happen they can be extremely costly: I already noted the record payments following the *Deepwater Horizon* (Macondo well) blowout and oil spill.

Myron Kinley put out the first well blowout with explosives in 1913, and a decade later he set up the first company to control wild wells. Kinley trained Paul Neal Adair, who formed the Red Adair Company in 1959 and became the most famous of all oil-well fire fighters, as well as Asger Hansen and Edward O. Matthews, who, in turn, left Adair's company in 1978 to set up another famous oil-well control outfit, Boots & Coots. In 1997 International Well Control (formed after Adair's retirement and the sale of his company) acquired Boots & Coots and reunited the top expertise in fast control and safe capping of runaway wells. Safety Boss in Calgary is another oil-well control company operating globally. Extinguishing more than 700 oil wells set on fire in Kuwait in March 1991 by the retreating Iraqi army was the biggest ever test of these skills.

Despite a high level of mechanization the operation of modern oil drilling rigs still requires a great deal of heavy (and often) dangerous labor during which most of the crew (now typically about twenty workers per land rig) is exposed to weather. The lowest-ranking members of a drilling crew are roustabouts who do all unskilled peripheral manual jobs ranging from offloading and moving the drilling supplies, stacking drill pipes and mixing drilling fluids to cleaning, scraping and painting the deck. Roughnecks handle all the manual tasks of the drilling operation, adding new lengths of drill pipe, disassembling the drill string prior to changing the drill bit or retrieving core samples and cleaning all drilling equipment. Skilled jobs are those of motormen, derrickmen, assistant drillers, and the driller who supervises the entire crew and operates the pumps, draw-works and rotary table from a console in his control room. Drilling usually proceeds around the clock, requiring two or three shifts on land and two twelve-hour shifts offshore.

Better drills and faster drilling operations combined with the need to search for oil in deeper reservoirs led to a steady increase of both record and average drilling depths. Record US well depths reached with rotary rigs increased from 300m in 1895 to more than 1.5km by 1916; the 3km mark was reached in 1930, the deepest pre-WWII well was 4.5km (in 1938) and the 6km mark was surpassed in 1950. Beyond this the drilling entered the ultradeep range where high temperatures and high pressures, and often a highly corrosive environment, present new challenges. Many of these wells were drilled in Oklahoma's Anadarko basin where the 9km barrier was passed in 1974. The average depth of new US exploratory oil wells increased from about 1,460m during the 1950s to nearly 2,300m during the first decade of the twenty-first century. The three most notable post-WWII advances in oil drilling were the move into deeper offshore waters, the development of directional drilling, and the emergence of horizontal drilling and hydraulic fracking as a fundamental new method of crude oil extraction.

Offshore drilling uses the same tools and procedures as operations on land but it faces major technical challenges in positioning the drilling platforms. For decades, offshore drilling was just a simple extension of onshore operations. The first drilling from wooden wharves extending short distances from the shore was done in California as early as 1897. The first permanent wooden production platforms were built in the shallow waters of Lake Maracaibo in 1924; by 1927 concrete pilings were providing better anchoring, and in 1934 came the first standardized steel platforms whose numbers eventually multiplied to dot the lake with tall metal structures.

The first near-shore well was drilled in the Gulf of Mexico in 1937 and by the mid-1940s Texas prospectors had rigs drilling from platforms. Baku's oil is only partly onshore and hence from the earliest days of its exploitation efforts were made to find the best solutions for drilling offshore. Between 1909 and 1932 Bibi-Heybat Bay was filled with earth in order to drill off-shore deposits; in 1924 a small woodpile island was built in Bayil Limani and in 1949 drilling began from a large steel offshore pier at Neft Daşlari. In 1947 the first oil well (Kermac 16) drilled out of sight of land (nearly 70km from the Louisiana shore) was completed in 6m of water by Kerr-McGee Corporation in the Gulf of Mexico. Offshore drilling diffused from the Gulf of Mexico to California, to the further reaches of Lake Maracaibo and to Brazil's coastal waters, and it progressed from small submersible units suitable only for shallow waters to a variety of drill ships and semisubmersible rigs designed for year-round drilling even in such stormy waters as the North Sea.

There are many marginal shallow oil-bearing formations in all old oil provinces that were bypassed during the original development that sought more substantial reservoirs. Short-radius (less than 20m) drilling to produce horizontal laterals from thousands of abandoned and existing wells could add to production without any new surface disturbance. During the late 1990s research began to develop lightweight, flexible composite drill pipes made of carbon fiber and epoxy resins to be used in such short-radius drilling. Several composite drilling applications, including 10m-long sections of flexible drill pipe and composite drill collars, are now commercially available. These pipes can also have cables or fiber optic links embedded in their walls allowing for an unprecedented degree of logging while drilling, creating a smart drill pipe.

Early drillers sank many wells that deviated significantly from the vertical; after all, a deviation of just 10° will take a well nearly 500m off the vertical at a depth of 3km. As a result, boreholes

OFFSHORE DRILLING

In 1949 John Hayward combined a submersible barge and a piled platform to build the first submersible rig, *Breton Rig 20*; he devised a way to stabilize a ballasted barge on the seafloor so that only the columns (connected to the barge and supporting the rig's working deck) were exposed to waves. Kerr-McGee bought the rig in 1950, after it had drilled nineteen wells. A better version, *Mr. Charlie*, was built in 1953 by Alden J. Laborde who founded the Ocean Drilling & Exploration Company (ODECO). That company eventually combined with three other pioneers of marine drilling to form Transocean, the world's leader in offshore engineering.

The first self-elevating (jack-up) drilling rig, *Offshore Rig 51*, began working in 1954. Its ten legs (1.8m in diameter and 48m long) ended in large spud cans to minimize the pressure of the ocean floor. A prototype of modern jack-ups – a triangular platform with three legs and pinions driven by electric motors – was built by Le Tourneau Company in 1956. Jack-ups were not suitable for deeper waters and in 1961 Shell Oil was the first company to deploy a semisubmersible rig, *Bluewater I*, in the Gulf of Mexico. Different designs followed in rapid succession – and a decade after the first model there were thirty semisubmersibles at work.

During the 1970s Transocean introduced the Discoverer-class drill ships whose operations proceeded to set repeated drilling records: by the year 2000 the fifth generation of these ships could drill in waters 3km deep. A new chapter of offshore drilling began with operations in very deep water. The industry defines work at a depth of 1,500m as ultra deep drilling, and the Gulf of Mexico has the world's highest concentration of such projects and associated records, including an exploratory well drilled in 2003 by Discoverer Deep Seas in more than 3,000m of water. In early 2016 the *Maersk Venturer*, a drillship working for a consortium of Total and ExxonMobil, set a new record by drilling in water depth of 3,400m offshore Uruguay. These efforts have been rewarded with regular new major deepwater finds in the Gulf of Mexico (both in US and Mexican waters), offshore Brazil (in fields lying beneath salt and rock deposits up to 5km thick 2km below the sea level) and Guyana, West Africa and Angola and, most recently, in South-East Asia. By the middle of 2017 there were about 820 mobile offshore rigs (jack-ups, semisubmersibles and drill ships) marketed worldwide, with about 11% of them in US waters.

in early oilfields tapped unintentionally into neighboring drilling claims; sometimes they even merged with another well. The significance of this unintentional deviation from the vertical axis was recognized as a problem for the first time during the mid-1920s in Oklahoma and it led eventually to the development of directional drilling (deviating by increasingly greater angles from the vertical) and then true horizontal drilling and extended reach drilling (where horizontal components are at least twice as long as the vertical bore).

The first short horizontal wells were drilled in 1929 in Texas and in 1937 in the USSR, and John Eastman and John Zublin designed and used the first short-radius (6–9m) drilling tools in California during the early 1940s. These trials confirmed the feasibility of this new technique, but its high costs did not justify commercial adoption and it was only during the late 1970s that interest in horizontal drilling was rekindled both in North America and in Europe, and the process did not become a commercial success until the 1980s. During that decade North American hydrocarbon companies drilled more than 300 horizontal wells (mostly with a medium-radius of around 30m), but ten times as many were completed during the 1990s, when an increasing number were drilled outside of North America. Initial doubts about not being able to drill beyond 70° inclination were soon dispelled with successful 90° wells. Small deviations could be effected by using different weights on a bit or by varying drill speed. Greater departures from the vertical were obtained by placing inclined hard steel wedges (whipstocks) on the bottom of a borehole (or on top of a temporary cement plug), forcing the drill bit to deviate in the desired direction; obviously, this allowed only limited directional control and led to many missed targets.

The solution was to harness the kinetic energy of the drilling fluid to power the drilling bits. By the mid-1950s turbine drilling was converting the hydraulic power of the drilling mud into rotary power turning the bit, but today's positive displacement motors deployed in directional drilling are reverse applications of René Moineau's concept of an irrigation pump that could tolerate a high influx of solids. Unlike the turbines with hundreds of rotor-stator power section stages, modern drilling motors have a small number of stator and rotor lobes ranging from single-lobe (one rotor) designs to motors with nine rotors and ten stators. Rotors are placed eccentrically and their motion about the stator axis acts as a gear reduction that lowers the speed and increases the torque as the number of lobes increases.

Positive displacement motors are powered by the pressurized drilling fluid (and consequently called mud motors) that is commonly delivered at rates of 300–1,000 liters per minute (with large-size motors providing more than 6,000 liters per minute). Their rotation is restricted mostly to between 80 and 250rpm, and the latest designs have a maximum power output of 400–600kW. These motors are used to point the bit away from the vertical axis when the drill string is not rotating and to drill in a new, non-vertical direction; once this new direction is well established it can be maintained by rotating the entire drill string, including the bent section. This is an inefficient arrangement (as extreme torque and drag restrict the drilling capability), and one that also produces an uneven (overgauge and spiral) borehole that may cause difficulties for logging and casing operations. These problems were eliminated by the introduction of rotary steerable drilling motors during the mid-1990s.

The latest models integrate a high-power drilling motor with continuous drill string rotation. The drill string rotates only at a speed sufficient to agitate and to remove the cuttings and its continuous motion also transfers weight to the bit more efficiently. But the drill string's rate of rotation is decoupled from that of a high-speed drill (powered by a mud motor via gearing) whose steering and control are energized from a power-generating module (a turbine-driven alternator) in the bottomhole assembly. This arrangement keeps the well bore clean, produces a smooth and in-gauge borehole and allows for a high rate of penetration. Steerable downhole motors are also used in coiled tubing drilling, introduced during the late 1990s. This method has replaced standard rigid steel pipes with narrow (2.5–11.25cm in diameter) flexible steel tubing that is wrapped on a large spool mounted on a heavy trailer and straightened before it is unreeled into a well bore. Overall lengths of continuous coiled tubing (unit winding capacity) reach more than 7,000m for the smallest diameters. Slim-hole drilling eliminates the laborious tripping, and new high-speed turbine drills (with up to 10,000rpm) produce very high rates of penetration.

The benefits of directional drilling are obvious: for example, if the oil-bearing strata are 3km below the surface, then drilling at 70° rather than at a 60° angle off the vertical will make it possible to reach from the same drilling site an area that is nearly 2.5 times larger. The most extreme case of directional drilling is following a near-horizontal (in excess of 80°) or a perfectly horizontal direction, or even an undulating pattern along it (see figure 16). Such wells have boosted recovery rates as they can be directed through relatively thin oil-bearing strata that would be uneconomical to drill with a series of vertical wells. The world's longest extended reach well by the end of the twentieth century was BP's Wytch Farm M11 in Dorset, completed in 1998 in just 173 days after reaching 10,658m (vertically only 1,605m) to tap deposits under Poole Bay. That record has been repeatedly broken, and as of early 2017 the longest producing well was the one drilled in April 2015 by Rosneft at Chayvo field, from a drilling platform offshore from Sakhalin Island: its record-breaking length was 13,500m with a horizontal reach of 12,033m.

Early horizontal wells could be two or three times more expensive than a vertical well of the same length (mostly because of problems with high torque and drag while drilling, and with the well bore's integrity) but through their superior penetration of an oil-bearing formation their production rate could be a multiple of that of conventional wells drilled in the same reservoir. As horizontal drilling became routine its costs declined rapidly and the resulting economic returns made it even more popular. As already noted, ten times as many horizontal wells were drilled in North America during the 1990s than during the 1980s, and the practice became routine in the first decade of the new century when it was combined with hydraulic fracturing. Horizontal wells also helped to open new production prospects in sandy reservoirs and in formations with water or gas problems or low permeability, but horizontal drilling brought a true revolution in American oil (and natural gas) extraction only once it

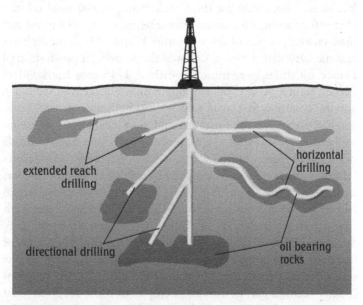

Figure 16 Directional and horizontal drilling

was combined with hydraulic fracturing: this process (as already noted) will be described in some detail in the fourth chapter.

History of oil discoveries

Many societies have known about hydrocarbons from seepages, bitumen pools and burning pillars, particularly encountered throughout the Middle East. But, unlike in the case of metallic ores, there were no attempts to discover underground oil deposits because the exploitation of heavier hydrocarbons was limited to a few (and localized) uses. Some ancient Mesopotamian cultures used asphalts and bitumens to inlay floor and wall mosaics and for protective coatings and burned lighter oils in fire pans for illumination. So did, centuries later, the Romans in some Asian localities (who, following the Greek example, also used oil as a charge for flaming containers during sieges and naval battles) and then various peoples of the Byzantine Empire, Muslim caliphates and the Ottoman Empire. Oil was also known in pre-industrial France: oil shales were mined as early as 1745 near Merkwiller-Pechelbronn in Alsace (oil was sold in small bottles as a medicine), and the country's first small refinery was built there in 1857.

The place with the longest tradition of local crude oil use (as well as small-scale exports) is undoubtedly the Absheron Peninsula of the Baku region on the Caspian Sea in Azerbaijan. Absheron's oil pools and wells were described by medieval Arabic travellers and historians, while a 35m-deep hand-dug well in Balakhani had an inscription dating it to 1593, and by the late eighteenth century there were scores of shallow wells from which oil (particularly the 'white' variety) was extracted for the production of kerosene (by primitive thermal distillation) for local lighting as well as for export (in skins) by camels and ships. In 1837 Russians (Baku became a part of the czarist empire in 1806) set up the first commercial oil-distilling factory in Balakhani, and in 1846 the world's first exploratory oil well was dug manually to a depth of 21m in Bibi-Heybat.

During the 1850s the rising cost of whale oil used for illumination led a number of entrepreneurs to search for alternatives and to the small-scale beginnings of new oil-based industries. In 1846 Abraham Gesner (Canadian physician and geologist) began to produce kerosene from coal and installed the first city lights in Halifax in 1850 and in 1854 set up the North American Kerosene Gas and Lighting Company on Long Island. Ignacy Łukasiewicz, a Polish pharmacist, invented the first kerosene lamp and he was also the first industrialist to distill kerosene from oil found in seeps near Krosno (at that time in the Austro-Hungarian Empire) where he opened an oil mine in 1854 and two years later the first distillery. America's first oil well was hand-dug in 1858 in a swamp near Black Creek hamlet in Lambton County, south-western Ontario, by Charles Tripp and James Miller Williams. Extraction was by buckets dipped into the well and emptied into barrels which were then transported for primitive refining in large, wood-heated cauldrons to yield kerosene and the residual lubricating grease. By 1859 the area was experiencing the world's first oil rush as burgeoning Black Creek was renamed Oil Springs.

In the same year came the first US oil rush. Oil seeps in western Pennsylvania were known by the Seneca tribe and during the eighteenth century bottled 'Seneca oil' was sold as a medicine. George Bissell, a New York lawyer, set up the Pennsylvania Rock Oil Company (later Seneca Oil) after Benjamin Silliman at Yale (hired by Bissell) confirmed that oil's distillation yields kerosene. In 1859 Edwin Drake (sent by Bissell to start regular production at Oil Creek, near Titusville, Pennsylvania) hired a local blacksmith to find the source of the largest seep by drilling the first US oil well. After a slow progress of about one meter a day the well struck oil (extracted by a hand-operated pump) at the depth of 21m on August 27, 1859. The well initially yielded twenty-five barrels a day, but by the year's end the flow was down to fifteen barrels.

With the market limited largely to lighting kerosene and lubricants the growth of new oil discoveries was relatively slow during the closing decades of the nineteenth century. The Canadian oil boom intensified in 1862 when Hugh Nixon Shaw, going against the prevailing wisdom that no oil would be found at depths below 20m, produced the world's first gusher from the depth of 60m, but soon the reservoir pressure began to decline and the Oil Springs wells had to use steam pumps. A second oil rush came with new discoveries at Petrolea (soon renamed Petrolia), about 15km north of Oil Springs, in 1865, but the Ontario oil boom was over by 1891 when John D. Rockefeller's Standard Oil Company flooded the market with cheaper American oil. The company was set up in 1870 and within a decade it controlled about 80% of the market for kerosene and other refined products.

In 1875 Ludvig and Robert Nobel (Alfred's brothers) launched their Caspian oil enterprise (which eventually became Baku's leading oil-producing enterprise: Nobel Brothers Petroleum Company) and in 1877 they built the world's first oil-carrying steamship, *Zoroaster*. In 1878 an oil gusher at Bibi-Heybat demonstrated the oilfield's exceptional size: the field eventually produced enough crude oil to be classified as the world's first known giant oilfield, that is, one with at least 500 million barrels of recoverable crude oil (see figure 17). The Rothschild brothers began investing in Baku in 1883 when they established the Caspian and Black Sea Oil Industry and Trade Society.

The first US giant oilfields were Bradford (1875) and Allegany (1879) in Pennsylvania and Brea-Olinda (1884) and McKittrick (1887) in California. The first oil-producing well in Texas was drilled in 1866 but the first economically important strike was in 1894 in Corsicana, made by a city crew drilling for water, and in 1895 the area's first producing wells were drilled by Joseph S. Cullinan who later became one of the founders of the Texas

Figure 17 A forest of wooden oil derricks in Baku in 1886

Company (which became Texaco, in 1901). The most important discovery came on January 10, 1901, when a well drilled by Anthony F. Lucas in Spindletop, near Beaumont, produced a 30m-tall gusher that took nine days to cap. Another oil boom followed, but Spindletop's enormous output (100,000bpd on the day of discovery, January 10, 1901; 17.5 million barrels in 1902) brought an oil glut and prices plummeted to three cents a barrel. Other notable pre-WWI oil discoveries were made in Romania, Indonesia, Burma and Iran.

As in the Baku region, pools and shallow wells of oil had been known in southern Romania for centuries, and during the 1840s workshops using simple distillation were built in Lucăcești. A true commercial refinery, using cast-iron cylinders, began operating in Ploiești (60km north of Bucharest) in 1857. The country's only giant oilfield, Moreni-Gura Ocniței, was discovered in 1900.

The discovery of oil in 1883 in northern Sumatra, at that time a part of the Dutch East Indies, led to the establishment of the *Koninklijke Nederlandsche Maatschappij tot Exploitatie van Petroleum-bronnen in Nederlandsch-Indië* (Royal Dutch Company for Exploration of Petroleum Sources in the Netherlands Indies) by Jean Baptiste August Kessler, Henri Deterding and Hugo Loudon in 1890. In 1907 the company merged with the Shell Transport and Trading Company set up in 1897 by Marcus and Samuel Samuel (in 1833 their father, Marcus, established a small shop in London selling seashells, a legacy that created one of the world's most recognizable logos) to form Royal Dutch Shell.

The Burmah Oil Company was set up in Glasgow in 1886 and it began producing oil a year later, with most of its extraction coming from moderately sized fields at Yenangyaung, Chauk and Minbu. Iranian exploration began with the concession granted in May 1901 to William Knox D'Arcy to drill in Ghasr-e Shirin and Chah-Sorkh. By 1904 D'Arcy had spent what was for that time the enormous sum of £220,000 without striking oil in commercial quantities. His enterprise was rescued in 1905 by new funds provided by the Burmah Oil Company. A new syndicate negotiated a contract (never recognized by the Persian government) with Bahktiari chiefs to drill in the south-west, near the Iraqi border, and on May 26, 1908 came the discovery of the first Middle Eastern oilfield at Masjid-e-Soleiman. In 1909 the Anglo-Persian Oil Company (the precursor of British Petroleum) was set up to develop the field and in 1914 it reached an agreement with the British Admiralty to supply oil for warships being converted from coal to liquid fuel (both the British and the German navies decided on the switch just before World War I) and the government became the company's majority shareholder.

During the early decades of the twentieth century, important strikes were made in North Texas, and in October 1930 the state's biggest field, East Texas, was discovered by a wildcatter, C. M. Joiner, who took up the claim after it was rejected by geologists working for large oil companies as unpromising. Another frenzied boom and another overproduction followed and in August 1931 the field was placed under martial law and the Texas Railroad Commission began to enforce a production quota. East Texas was the largest of the three Texas giants that began production between 1925 and 1950: the other two were Yates, discovered in 1926, and Wasson, in West Texas near the New Mexico border, in 1936. The latter field was exceptionally large (about 25,000 hectares) and by the end of the 1990s its cumulative extraction was surpassed only by the East Texas oilfield.

Exploratory drilling began to expand just before the century's end and in retrospect it is clear that the entire first half of the twentieth century was the golden age of oil discoveries as news of new giant oilfields became commonplace. In 1900 the US had only seven giant oilfields, but by 1925 the total had risen to seventy-five, and by 1950 it was 220. Two Californian fields were the largest American oilfields discovered between 1900 and 1925, Midway Sunset in 1901 and Wilmington Trend in 1922. Mexican crude oil production began in 1901 and by 1920 the country was the world's second largest producer, and the top exporter, of crude oil. This position was lost despite the discovery of Poza Rica field near Veracruz in 1932, because of major Venezuelan finds and even more impressive Middle Eastern oil discoveries. The first Venezuelan giant, Mene Grande on Lake Maracaibo's east coast, was discovered in 1914, while the supergiant Bolivar Coastal was added in 1917, and two giants, La Paz and Quiriquire, during the 1920s.

The first major Middle Eastern strike outside Iran was made by Turkish Petroleum, set up in 1912 to seek concessions from the Ottoman government to drill in Iraq. The deal was finally concluded with a new Iraqi government in 1925, and oil was discovered at Baba Gurgur, just north of Kirkūk, in October 1927. Production by the (renamed) Iraq Petroleum Company began in 1934 and Kirkūk proved to be a supergiant that produced the bulk of Iraqi oil until the Rumaila field near Basra (discovered in 1953) entered production. In Iran exploration by the Anglo-Iranian Oil Company added giants at Gachsaran and Haft Kel in 1928, Naft-i-Said in 1935, Pazaran in 1937 and Agha Jari a year later.

On May 29, 1933, the most important concession to date to explore for oil was signed in Jiddah by the king 'Abd al-'Azīz (the head of the newly constituted [in September 1932] state of Saudi Arabia). California Arabian Standard Oil Company (CASOC), an affiliate of Standard Oil of California (Socal, today's Chevron), made an initial payment of £35,000 in gold for the rights to prospect for oil in al-Hasa (today's Eastern province). Prospecting began in September 1933 and the first giant oil-field, Dammam on the western shore of the Persian Gulf near Dhahrān, was discovered in 1938. This find was soon followed by discoveries of the nearby (and much larger) Abqaiq and Abu Hadriya and Qatif (just north of Dhahrān). In 1944 CASOC became Arabian-American Oil Company (Aramco), and in 1948 the company discovered al-Ghawār, south-west of Dhahrān, that was, after extensive drilling, confirmed in 1956 to be the world's largest accumulation of crude oil.

During the late 1940s came the discoveries of the first Canadian giants in Alberta, Leduc-Woodland in 1947 and Redwater in 1948. Russia's, and after 1921 the USSR's, oil extraction was dominated by the Caspian oilfields until after World War II. Three giant fields were discovered in the Baku area, Balakhani-Sabunchi-Romani in 1896, Karachukhur-Zykh in 1928 and Neft Daşlari in 1949. After World War II the region between the Volga and Ural rivers rapidly emerged as the country's largest oil province, with giant fields discovered in Tuymazy (in what is now autonomous Bashkortostan) in 1937, Mukhanovo in 1945 and Romashkino (in today's Tatarstan) in 1948.

The enormous post-WWII increase in demand (detailed in chapter 1) was easily met by new giant oilfield discoveries during the 1950s and 1960s, mostly in the Middle East. The most remarkable finds of those two decades included the supergiants in the Persian–Zagros oil province. The USSR's center of oil production began its second shift, from the Volga-Ural region to Western Siberia, with the discovery of the supergiant Samotlor (in 1965), and Prudhoe Bay on the North Slope of Alaska (1968) became America's largest supergiant.

Canadian reserves were boosted by giants at Pembina and Weyburn-Midale, Swan Hills and Judy Creek, all found in the 1950s. New entries to the list of countries with major oil reserves included Algeria, Libya and Nigeria. And in 1959 Russian and Chinese geologists discovered the supergiant Daqing oilfield in Heilongjiang. This field produced about 50Mt of oil every year for twenty-seven consecutive years and 33Mt for twelve years, an uncommon productivity level, and nearly six decades later it is still China's largest oil producer, in 2015 about 25% ahead of off-shore Bohai field. The first major offshore finds in the relatively shallow waters of the Gulf of Mexico and the Persian Gulf also took place during the 1950s and early 1960s.

During the 1970s – the decade of unprecedented OPEC-driven oil price rises, ensuing 'energy crises' (see chapter 1 for details) and global economic slow-down – the discoveries of giant oilfields were still (albeit only marginally) ahead of the total for the 1960s, but the total amount of recoverable reserves in these fields was nearly 50% lower: new-found giants were getting, on average, noticeably smaller. The biggest addition was Mexico's Cantarell Complex in 1976 that eventually became the world's third largest field in terms of average daily production, and North Sea exploration was finally rewarded with several giant finds, all in Norwegian waters.

After the collapse of high oil prices (in 1985) global hydro-carbon exploration shrank dramatically while the price fluctuated but remained (in inflation-adjusted terms) well below the 1981 peak. Investment in new exploratory activities remained low during the 1990s and it began to rise significantly only with the rise in oil prices in 2005. Nevertheless, the last two decades of the twentieth century brought some notable giant oilfield discoveries; including Norway's Draugen (1984), Heidrun (1985) and Norne (1992), Mexico's Caan and Chuc in 1985 and Sihil in 1999, Iran's supergiant Azadegan (in 1999), Kazakhstan's supergiant Kashgan (2000), Brazil's Marlim (1985), Albacora (1986) and Roncador (1996) and new US giants, Ursa (1991), Auger (1996) and Alpine and Thunder Horse (1999).

But, unmistakably, the rate of new giant oil discoveries was declining after peaking at nearly 180 during the 1960s: by the 1980s it was down to about 80, and during the 1990s it reached only 37 (or 43, depending on the reserve estimates), while the total of estimated reserves present in these newly discovered fields declined even faster, from the peak of about 270Gb during the 1960s to only about 30Gb during the 1990s (but, as is almost always the case, reserve appreciation has already raised the latter total). These were undesirable trends because (as explained in chapter 2) the frequency of the world's oilfield sizes is hyperbolically distributed. But the discovery rate of oil giants was significantly up between 2000 and 2009, and the second decade of the twenty-first century is on track to see the fourth highest decadal gain since 1868. Given the skewed distribution of oilfields it is inevitable that the countries with the highest number of giants (or with a few supergiants) are the ones with the largest oil reserves. In 2018 the top five countries on this list were (with shares of the world's conventional oil reserves, rounded to the nearest percent in parentheses) Saudi Arabia (20), Iran (12), Iraq (11) and Russia and Kuwait (each with nearly 8% of the total). The addition of nonconventional oil resources scrambles this order (see chapter 5).

Another remarkable fact is the astonishing growth of global oil reserves during the second half of the twentieth century, a progress that has been recorded since 1945 thanks to the worldwide surveys conducted by two of the oil industry's leading journals, *World Oil* and *Oil & Gas Journal*. A retrospective look using *Oil &*

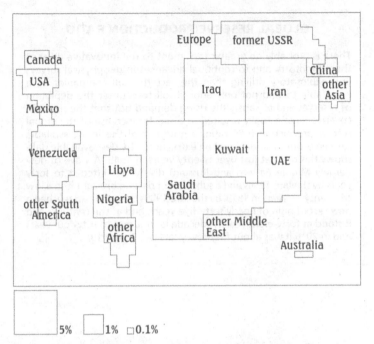

Figure 18 Map of the world with the size of countries and continents proportional to their share of conventional oil reserves in 2015

Gas Journal estimates shows the growth of global oil reserves from just 50Gb in 1945 and 85Gb in 1950 to about 715Gb by 1974: no other period of twenty-five years has seen, or is ever likely to record, a similarly huge gain. After a decade of stagnation global oil reserves surpassed 900Gb in 1988, the 1Tb mark (1.007Tb) was reached in 1995 and by the year 2000 the total had risen to 1.028Tb. And then, thanks to the inclusion of nonconventional reserves (primarily in Venezuela and Canada) that were previously excluded from annual summaries, the global total rose to 1.292Tb in 2005, and in 2017 it stood at 1.7Tb.

GLOBAL RESERVE/PRODUCTION RATIO

There is probably no greater testament to the innovative drive of the oil industry and to continual advances in geophysical surveys and exploratory drilling than the fact that oil companies have been able to find not only enough oil to counter the depletion of reserves and to satisfy the rising demand but that the new discoveries have actually resulted in a secular increase of the global reserve/production (R/P) ratio, a quotient of the latest available reserve estimate and annual oil extraction. The best available data shows this ratio at just over twenty years in 1945. A wave of new (largely Middle Eastern and Russian) discoveries lifted it to forty years by the late 1950s and a subsequent decline brought it to a low of twenty-six years in 1979. By the late 1980s the ratio had reached a new record high of nearly forty-five years, and at the century's end it stood at forty-eight years. A decade later it rose to fifty-four years and in 2016 it was about fifty-one years (see figure 19).

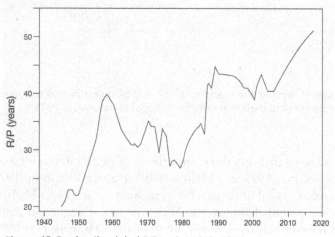

Figure 19 Crude oil's global R/P ratio, 1945–2015

Preventing the global R/P ratio from falling to uncomfortably low levels would still count as a remarkable achievement but that new exploration and development were able to boost the ratio to more than fifty years despite the 24-fold rise of global oil

GLOBAL RESERVE/PRODUCTION RATIO (*cont.*)

production between 1950 and 2017 is one of the most astonishing, if little appreciated, achievements of the modern oil industry. While the overall trend of rising global R/P ratios is real, there are doubts concerning some of the changes in the reported totals of national oil reserves. This concern is due to the absence of rigorous and uniform international standards; as a result, many values are not comparable and some national totals have always been suspect. Proved reserves, the concept used by the US Securities and Exchange Commission (SEC), refer only to the oil that is ready to be produced in the near term. Proved and probable reserves (used in Canada) also count oil where the probability of production over the lifetime of an oilfield is fairly high. And there are estimates of possible reserves for which the probability of eventual extraction cannot be realistically appraised.

This chronic incompatibility problem was overshadowed in 1987 when the record increase in world oil reserves took place not because of any exceptional discoveries (at that time worldwide exploratory activity was actually at its lowest post-1950 level) but because six OPEC members simply announced a massive upward revision of their oil reserves. This upgrade accounted for nearly 90% of the overall 27% jump recorded in global oil reserves in 1987. Iraq and Iran, at that time at war, made the largest revisions of, respectively, 112% and 90%. Upward revaluations of existing reserves are the norm in the oil industry but the magnitude of this increase and its timing (following the collapse of world oil prices) made it suspect. Other suspect totals have remained unchanged for years.

Only the opening of books by national oil companies (an unlikely prospect) would settle the matter. At the same time, oil reserve claims that must be filed with the US Security and Exchange Commission by all publicly listed oil companies are definitely on the conservative side, and the Russian oil reserves (Russia uses its own resource classification developed during the Soviet era) are almost certainly higher than the published Western estimates. On the other hand, we may see further additions to several national totals as technical advances in extraction will transfer parts of their resources of nonconventional (heavy, shale) oil to reserve category. In any case, R/P ratios should not be seen as a highly accurate accounting tool but as a convenient and revealing indicator of commercially available oil resources that also informs us about the prospects of future extraction.

4

How oil is produced, transported and processed

The discovery of exploitable amounts of crude oil is only the beginning of a long process of developing a field, managing extraction, transporting and refining the oil and distributing the refined products to their final markets. Proper field management aims at sustaining the longest possible period of production; some currently productive fields are now more than 100 years old.

Once a borehole is drilled to the desired depth the results of logging must be carefully evaluated in order to decide if the well is to be completed; that is, if the borehole will be protected with the heavy steel casing that is cemented in place to create a well, or if it will be abandoned (plugged with cement). Exploration is a success if there are sufficient indications that hydrocarbons are present in commercially recoverable concentrations; in the oil industry's argot, a prospect (also called an anomaly) turns into a play (an area where hydrocarbons are known to be present). But even major finds are not converted immediately into production wells. Their exploitation must await the additional drilling that is required to delineate a field in greater detail, to determine the volume of recoverable oil more accurately and to decide on the best course of the entire field's development. And even then a decision can be made to temporarily shut down the wells if falling oil prices turn the development of a new and relatively expensive oilfield into a marginal proposition. On the other hand, old shut-down marginal wells that require expensive pumping may be reopened if oil prices rise.

The high concentration of oil production in a handful of the largest oil provinces and the rising worldwide consumption of liquid fuels had to be bridged by the emergence of large-scale transport and storage techniques. Worldwide shipments of crude oils and refined oil products (over long distances by pipelines and tankers, or shorter distances by railway tank cars and trucks) add up to the single largest transfer of a mass-produced mineral commodity on Earth.

Crude oil is such a complex mixture of organic compounds that its direct combustion would represent an enormous waste of a valuable non-renewable resource. Separating the mixture into several principal categories with a more homogeneous composition (but still of considerable heterogeneity) adds a great deal of value to the final products that can be marketed for specific uses at prices commensurate with their relative scarcity and overall utility. This process of separating crude oil into specific fuels (and non-fuel products) is done by refining, that is, by subjecting crude oil to a variety of physical and chemical treatments designed to maximize the yield of the desired final products.

Oil production

Success in exploration leads to more drilling before a reservoir is developed for commercial production. More wells are sunk in order to delineate a reservoir's extent and depth and to emplace production wells in order to optimize long-term extraction. During the subsequent decades new wells are drilled to monitor, maintain and to enhance production (surveillance and injection wells). As oilfields age and reservoir pressure drops, new wells are sunk in order to tap previously unexploited parts of a formation, to enhance oil recovery by introducing fluids into oil-bearing strata through special injection wells, or to store gas. Old bore-holes are re-entered to deepen the existing wells and both active and abandoned boreholes can be used as new entries for hori-zontal or multilateral drilling. Development wells are thus much more numerous than exploratory ones, and well completion aims at ensuring that oil can keep flowing into a well and that it can be reliably brought to the surface while water (be it above or below the oil-bearing strata) and the rock formation surrounding the well are kept out.

Modern wells are completed with steel casing extending over the entire oil-bearing zone (the casing is run all the way to the well's bottom where it is cemented in place) and hence it is necessary to ensure that oil will flow into the well from the surrounding formation. Various mechanical methods have been used to puncture holes in the casing, including firing special bullets from short-barrel guns. The standard practice today is to use shaped charges whose high-velocity detonation easily penetrates the well's steel casing and the surrounding cement and blasts short narrow holes into the oil-bearing rocks, allowing the free flow of hydrocarbons into the well. Perforations can be made in precisely placed spots along the casing in order to minimize the ingress of unwanted fluids (above all, salty formation water).

Most of the oil produced from deep wells is drawn through tubing, a special string of pipes inside the casing. The principal reason for this practice is the ease of repairs: unlike the casing this production tubing is not cemented in place (it is usually anchored in place just above the production zone by gripping elements that are latched to the casing's inside wall and by special rubber seals) and hence a joint or a section failure is easily repaired. In addition, it is easy to incorporate various flow control devices and automatic shut-off safety valves in the tubing and the casing remains protected from corrosion and erosion caused by hydrocarbons, water and dispersed solids drawn from a well.

A well is completed with the installation of the above ground wellhead.

This complex assembly stack of valves, spools, pressure gauges and chokes is commonly called a Christmas tree, a tell-tale sign of the presence of an oil well in an otherwise undisturbed terrain: it was introduced in 1922 and it is connected to the top of a completed well in order to control the flow of oil (see figure 20). A small-diameter gathering pipeline leads from a well to a stock tank and the entire area used for exploratory drilling can be returned to its previous use.

New wells are commonly under enough pressure (due to accumulated gases, water or gravity drainage) to push the oil from the reservoir into the well bore and then all the way to the surface (i.e. the reservoir pressure is higher than the pressure inside the well bore). In reservoirs where pressure is higher than the bubble point pressure, the undersaturated oil will contain variable volumes of solution gas, with the typical formation volume factor (ratios of reservoir/stock tank barrels) between 1.2 and 1.6. Oil volume thus shrinks as it reaches the surface but the expanding gas helps a well to flow naturally. The best natural flows come with the presence of cap gas when up to half of the oil in place (but more typically 25–35%) can reach the surface unaided. Natural waterdrive, with underground water displacing oil, is less efficacious, with free-flowing recovery rates of as little as 10% and, rarely, up to 40%. As the extraction proceeds the reservoir pressure declines (sometimes rapidly, in a matter of months, but usually only after many years) and it is necessary to deploy artificial lift. However, some reservoirs, especially those containing heavy oil, do not have sufficient initial pressure and oil production requires pumping from the very beginning.

Production under natural pressure or aided by a pump lift is classed as primary oil recovery and in most reservoirs it is able to extract only a small proportion of the oil that was originally

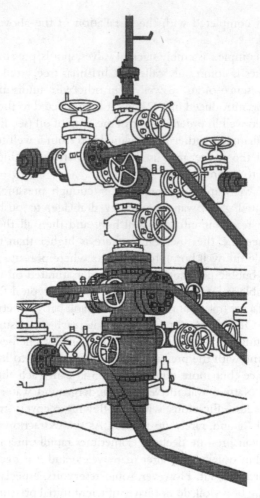

Figure 20 A 'Christmas tree' wellhead assembly (courtesy of BP)

present in the formation. The early producers followed an entirely primitive sequence of oil extraction: initially they were producing as much oil as a reservoir's natural pressure allowed (and all too often drilling too many wells to do so). Once the natural flow ceased they deployed pumps and once the pumping became uneconomical they considered a reservoir depleted and moved on. Natural gas associated with oil was simply released into the atmosphere or it was flared, that is, burned in the open. This wasteful practice has been greatly reduced worldwide, but it continues in some regions with large flames recorded as bright lights by Earth-observation satellites.

LIFTING OIL

External energy is needed to lift the fluid in a well bore. This has been done using rod pumps, electrical submersible pumps and also by gas lift. Rod (beam) pumps are readily recognizable because of their characteristic massive assembly – a walking beam that is moved by a pitman attached to a crank and counterweight usually powered by a diesel engine – and because of their ponderous nodding motion.

An electrical submersible pump was used for the first time in Baku in 1917 but the practice spread only after Phillips Petroleum introduced better designs during the 1920s. Modern designs include pumps with longer strokes and slower speeds that result in more complete pump filling and in higher pumping efficiency; a low-profile unit that can work under center-pivot irrigation pipes in crop fields; and an entirely new design that replaces rods with computer-controlled cables. Modern submersibles are multistage centrifugal pumps that are installed just below the fluid in a well bore and electrified via an armored cable. Subsurface hydraulic pumps use power fluid that flows down an inner tubing string; the return flow of this fluid, mixed with the produced oil, goes through the annulus between the tubing strings. Where gas is readily available it can be used to lower the mass of the liquid column (gas lift systems appeared for the first time during the 1930s). More than half of the world's operating oil wells now use some kind of artificial lift, with sucker-rod pumps dominant.

Reservoir engineering, now a key pillar of profitable oil extraction, began to evolve slowly during the 1920s and it progressed more rapidly during the 1930s and 1940s. It has been revolutionized by the deployment of directional and horizontal drilling, by multilateral wells and by new methods of secondary recovery and enhanced production. As already noted (in chapter 3), horizontal drilling diffused rapidly during the late 1980s and 1990s. In addition to the already noted advantages (above all, increased production and recovery rates) there are other appealing attributes of the technique: it has made it possible to complete several wells from a single location, a particularly welcome capability when producing from offshore fields (where a single platform may gather oil from many wells), in land locations that are difficult to access and in situations where a number of smaller reservoirs can be reached from a single wellhead; it reduces the number of wells (or, even more importantly, of offshore platforms) needed to develop the reservoir; and it makes it possible to position wellheads away from environmentally sensitive locations and to develop near-shore reservoirs from land.

Reservoir development through the use of multilateral wells was first advocated in 1949 by Alexander M. Grigoryan. In 1953 in Ishimbai field (in today's Bashkortostan in the Russian Federation), Grigoryan used downhole turbodrills (i.e. operating without rotating drill strings) to drill the main 575m-deep wellbore that ended just above the pay zone, and from that point he drilled nine branches reaching into the production zone, one with a maximum horizontal reach of 136m. The world's first multilateral well cost about one and a half times as much as a vertical equivalent but it resulted in a seventeen-fold increase in production, a return that led to an additional 110 such wells being drilled in the field. But because of the significantly higher costs of multilateral drilling the technique's early successes were not followed by widespread adoption. Commercial diffusion of multilateral drilling has lagged about a decade behind the acceptance of horizontal drilling, with increasing numbers of such completions since 1995. The simplest multilaterals are open-hole bores from an open-hole mother bore and they are possible only in consolidated formations. A more common arrangement has open laterals running from the cased and cemented main bore. Multilateral wells drilled using mud motors guided by precision geosteering (sometimes done in a fishbone pattern radiating from the main well bore) make it possible to extend the life of aging oilfields and to achieve optimal reservoir drainage in new finds. Undulating well bores, following a roller-coaster oil-bearing formation are now also possible.

Intelligent well systems are yet another innovative technique that has acquired the status of a mainstream solution since the year 2000. The first system with permanent monitoring and the ability to control the subsurface flows was installed in 1997 and by 2005 there were more than 200 such installations. Intelligent well systems are aimed at maximizing output and reserve recovery rates by installing sensors, monitoring equipment and completion components (above all control valves, usually as either hydraulically operated or electronically actuated sliding sleeves) that are remotely controlled from the surface; their constant monitoring makes it possible to control, in real time, the flow of water or gas encroachment in producing wells as well as regulating the fluxes in water or gas injection wells. Data from intelligent well systems is used for increasingly complex dynamic simulations of reservoirs. These simulations can help to optimize the extraction rates and to anticipate the reservoir flows and well performance.

Given the cost of such installations they were originally best suited for high-cost (that is mostly deepwater) wells, for managing the production from multilateral wells and for boosting productivity in old wells in mature oilfields (by better managing the process of secondary and tertiary oil recovery). Intelligent wells are also used to develop marginal fields and to produce oil from heavy oil deposits. The North Sea and the Gulf of Mexico were the two oil provinces pioneering the installation of smart wells with the highest number of intelligent wells but such wells are now common in many onshore operations. The next logical step is to extend these capabilities and to develop entire intelligent oilfields.

SECONDARY OIL RECOVERY

The main purpose of secondary oil recovery is to restore adequate reservoir pressure and to move oil toward the well bore. This is most commonly achieved with water flooding. Unintentional flooding has been taking place since the earliest decades of oil extraction but controlled flooding in water injection wells has been used since the 1920s. However, the effect may be less than expected due to the variable permeability of the injected formation, and injected water may also break through into a well causing production and processing complications. Oil brought to the surface must be separated from water and from any impurities as well as from gases. The process reaches its limit once the injected fluid constitutes such a high share of the total produced volume that extraction becomes uneconomical.

In large fields secondary recovery means handling huge volumes of water. Al-Ghawār has been producing with the assistance of peripheral water flooding since the early 1960s, and by 2003 the water saturation of extracted oil had reached 33%. The seawater used in the process is delivered by a pipeline from the Gulf and bottom water is discarded, although further treatment may be required to separate water that has become emulsified in the oil. Water can also be mixed with polymers to enhance its viscosity and to lower its mobility. Surfactants can be introduced to gather oil and to surround it with a barrier, making it easier for the water to push it to the surface. Steam flooding, consisting either of a continuous injection of hot steam or of a cyclical introduction of steam and water into a reservoir, is common in recovering heavy oil.

Immiscible gas injection is a process replicating the natural gas cap drive and it has been used since the 1920s. More common, and much more efficient, miscible gas injection (used since the 1950s) introduces supercritical compressed CO_2 into the gas cap. Field management now routinely relies on optimally distributed gas-injection wells to enhance reservoir pressure. Other possible methods

SECONDARY OIL RECOVERY (*cont.*)

of improving oil recovery include microemulsion flooding (very expensive); *in situ* combustion whereby heat and steam created by the burning of bitumen is used to drive oil from a formation (rarely used and risky); and microbial recovery, that is, inoculation with suitable bacteria to produce biosurfactants that act as emulsifiers or wetting agents. Combined primary and secondary recovery can now extract as much as 40–50% of the oil initially in place, that is, 20–30% more than in operations without enhanced recovery.

By far the most consequential innovation in the early twenty-first century's oil production has been the combination of horizontal drilling and hydraulic fracturing that made it possible to produce crude oil and natural gas locked in shales. As already explained, horizontal drilling became a widely adopted commercial practice by the 1990s – and so was hydraulic fracturing, commonly known as fracking. Its distant precursor was oil-well 'shooting' (using nitroglycerin to shatter rocks) that was practiced during the early decades of American oil extraction to boost oil flow. Non-explosive procedures relying on acids were introduced during the 1930s and the first attempt at hydraulic fracking was made in 1947 by Standard Oil of Indiana in Kansas: pressurized fluid was pumped underground to fracture oil-bearing rocks, and fluid-borne solids (propping agents, or propants) held the fissures open to aid the release of hydrocarbons previously bound in a formation. The Halliburton Oil Well Cementing Company patented the process (calling it Hydrafrac) in 1949 and in March of that year completed the world's first two fracking treatments of oil wells in Oklahoma and Texas. The treatment boosted well productivity by an average of 75%; it spread widely during the 1950s and by 2010 the American oil and gas industry had used it about 2.5 million times for more than half of all completed wells.

But it was only in 1991 that George P. Mitchell combined the two techniques and the Gas Research Institute subsidized the first fracked horizontal well drilled by Mitchell Energy & Development. Six years later the company developed a cheaper fracking fluid containing sand and guar bean powder as a gelling agent, and in 2002 it was acquired by Devon Energy, which promptly combined its extensive experience in horizontal drilling with water-based hydraulic fracturing. What came next was an impressive exponential take-off of a highly effective technical innovation that has transformed America's crude oil and natural gas production and made the country, once again, the world leader in the extraction of hydrocarbons. In 2000 about 23,000 hydraulically fractured wells produced less than 2% of all US crude oil; by 2015 there were some 300,000 fracked wells supplying 51% of the country's total oil extraction.

But the fracking revolution has been accompanied by concerns about its social and environmental impacts. The influx of (mostly single male) workers into booming towns in regions containing the best shale deposits (North Dakota, Texas) has created housing and service labor shortages and other problems associated with rapid population growth, but worries about air and water pollution have attracted most attention. An enormous increase in heavy truck traffic on rural roads (with hundreds of tanker truckloads of fracking liquids required for completing a single well, and thousands needed for multi-well pad sites) and the ensuing noise and air pollution have been the common annoyance. But the risks of water pollution (contaminating aquifers, including drinking water supplies, with fracking fluids) and air pollution (emissions of methane from fracking wells often located fairly close to homes or schools) have been far more worrisome and they generated a great deal of concern among affected populations and attracted environmental activists seeking a new high-profile cause. Many lawsuits were filed and as a precaution some jurisdictions (counties, cities, states, most notably New York) banned fracking altogether or enacted temporary moratoria.

Fracking fluid is about 90% water. Most of the rest is sand, and additives (hundreds of substances have been tried) usually make up less than 0.5% of the volume but they contain a mix of chemicals (acids, corrosion inhibitors, gelling agents, surfactants, biocides) that should never be allowed to contaminate drinking water. Usually this is not a problem as fracking takes place far below the aquifers, and steel and cement in properly finished wells should prevent any contamination closer to the surface. But given the large numbers of fracking wells completed in short periods of time it is not surprising that some water supplies have been contaminated. The same is true of the presence of methane in drinking water, particularly in homes close to fracking wells used to produce natural gas. In addition, only a small amount of water used in fracking returns to the surface where it cannot be recycled infinitely (salt concentrations get too high) and has to be injected into deep wells, far below the aquifers and hence is removed from rapid water cycle. Atmospheric emissions associated with fracking range from the release of greenhouse gases (CO_2 and CH_4) to regional impacts of nitrogen oxides (contributing to the formation of ozone) and to local exposure to particulates from diesel exhaust and hydrogen sulfide and toxic volatile hydrocarbons (above all benzene, toluene, ethylbenzene, and xylene).

The challenge is similar to many other environmental problems in the past. To dismiss these concerns as unwarranted runs against the best scientific evidence: when improperly executed, hydraulic fracturing can be more than merely unpleasant and temporarily disruptive. That is why, as we accumulate more information about the extent of air and water contamination caused by fracking, we can assess the true extent of these challenges, bring in the requisite regulatory measures and deploy better operating methods (including the use of safer fracking fluids) and monitoring procedures. As in so many cases, public acceptance of hydraulic fracking rests not only on the best possible measures to minimize all environmental impacts but also on the rational assessment of risks: fracking has been, so far, confined mostly to the US, and the long-term health of the population and risks of sudden death will be affected far more by, respectively, the epidemic of obesity (with two-thirds of adults now either overweight or obese) and driving under the influence of alcohol and drugs.

Completing wells under the sea is obviously much more challenging than equivalent operations on land. Flow control devices can be placed above the water line in shallow waters but deepwater drilling has required the development of new subsea production systems. Completed undersea wells (with Christmas tree exposed to the surrounding water, sheltered in housing or buried in the seabed) are connected to manifolds on the ocean floor and tied with electrical cables and flow lines to shallower waters. The new hardware and techniques needed for these completions (subsea trees, remote robotic installations and repairs) were pioneered in the Gulf of Mexico, where the first submerged wellhead was installed in 16m of water in 1961.

The 1960s saw nearly seventy subsea completions (with wells connected to fixed platforms in waters up to 190m deep) as well as the first multiple-zone completions tying a number of wells to a single production facility. Increased drilling activity during the 1970s brought the first subsea tree systems that were installed entirely below the seabed and the first applications of diverless installation and maintenance of deepwater production sites (in four wells in the Ekofisk field in the North Sea in 1971). The North Sea was also the site of the world's first floating production system, set up at the Argyll oilfield in 1975, and subsequently many similar arrangements have been used at other oil and gas fields in the region as well as offshore Australia and Brazil.

Operations in waters deeper than 450m could not use fixed production platforms and the increasing volumes of offshore extractions have required larger production facilities; by the end of the twentieth century these structures included the most massive mobile artefacts ever built. In the 1970s the world's tallest and heaviest drilling and production platform was Shell's *Cognac* in the Gulf of Mexico: at 308m it exceeded the height of the Empire State Building. In November 1982, *Statfjord B* platform was towed from a construction dock to the Norwegian sector of the North Sea: its four massive hollow concrete columns and storage tanks at its base made it, at just over 800,000t, the heaviest object ever moved.

In 1989 Shell's *Bullwinkle*, sited in 406m of water and weighing about 70,000t, became the world's tallest pile-supported fixed steel platform. In 1983 the first tension-leg platform (TLP) in the Hutton field in the UK sector of the North Sea was anchored by slender steel tubes to the sea floor 146m below the surface. In 1998 *Petronius*, working for Chevron and Marathon in the Gulf of Mexico, became the world's tallest freestanding structure, although only 75 of its 610m were above water. The next year *Ursa* TLP, a joint project of a group of companies led by the Shell Exploration & Production Company, was the largest structure of its kind. With total displacement of about 88,000t (more than a *Nimitz*-class nuclear aircraft carrier), it rises 146m above water level and is anchored with sixteen steel tendons to massive (340t) piles placed into the seafloor 1,140m below the surface (see figure 21). But even before *Ursa* was finished *Hibernia* (in the Atlantic, offshore of Newfoundland) became the heaviest gravity-based structure: its maximum oil storage and water ballast add up to 1.2Mt. New records were set in the second decade of the twenty-first century. In 2010, the floating drilling and production platform *Perdido*, working for Shell in the Gulf of Mexico, began to operate in a water depth of 2,450m, gathering oil by umbilicals from 22 wells; in 2014, the Russian *Berkut*, offshore of Sakhalin, set the record for the heaviest structure with 200,000t (and able to withstand the strongest earthquake), and in 2016

Figure 21 *Ursa* oil production platform in the Gulf of Mexico

Shell's *Stones*, a drilling, storage and offloading floating platform in the Gulf of Mexico, began to operate in water 2,900m deep.

An accurate global count of all operating oil wells is not easy to establish: published counts often refer to all (oil and gas) wells, but the grand total should contain associated gas wells (that is, oil wells with some natural gas production) as well as stripper wells (low productivity wells that are the first ones to be shut down when prices decline). The most reliable revision of the global inventory shows 1.23 million wells operating worldwide in 2012, and that year's average daily extraction of 86Mbpd would pro-rate to about 70 barrels of oil per well a day. But that is a perfect example of a misleading average, both on the international and intra-national level. The already noted highly skewed distribution of oilfield sizes (see chapter 2) must also apply to individual wells. In 2012 nearly half of the world's operating oil wells were in the US; in 2014 their total surpassed, once again, 600,000 (between

1950 and 2010 it fluctuated between 500,000 and 650,000) with nearly two-thirds of them being low-productivity stripper wells. Historical US statistics show that average well productivity during the post-WWII era peaked in 1972 at 18.6bpd/well and that it has been between 10 and 11bpd/well since the late 1990s. In contrast, OPEC's 2015 average was about 700bpd/well, with Iran at about 1,300, Kuwait at just over 2,000 and Saudi Arabia at nearly 2,900bpd/well.

Given the enormous range of geological settings (location, stratigraphy, oil quality) and extraction methods, it must be expected that the cost of finding and developing oil deposits as well as the expense of producing oil (commonly called lifting costs) vary widely, with the extremes between the least and most expensive operations ranging by more than an order of magnitude. I will quote a number of representative values to indicate these cost ranges. As expected, supergiant and giant oilfields in the Persian Gulf region have the lowest lifting costs: some wells produce oil at less than 50 cents a barrel. In 1999 the Saudi oil minister quoted a nationwide mean of less than $1.50/b, while in 2004 a former secretary general of OPEC wrote that in a highly competitive world the organization could produce and sell oil to satisfy the global demand at $5/b. In 2005 the late Morris Adelman calculated the average cost of a Saudi barrel at post-2005 prices at $2.90 – but cautioned that the value was too high as it included the associated natural gas whose value cannot be separated due to the absence of adequate data. The best recent estimates put Aramco's average lifting still at less than $5/b.

Average drilling and completion costs of America's onshore oil wells have declined from the peak reached in 2012 and they are now mostly between $5 million and $7 million/well, with the total split about 1:2 between drilling and completion. The cost of finding and developing new reserves has a particularly wide range, from less than 10 cents a barrel for oil sands to $25/b for deep water discoveries, while production (operating) costs may range from just a few dollars per barrel in the richest Middle Eastern oilfields to $30–35 for offshore wells in deep water. Common US costs for finding and developments have been recently between $20 and $25/b onshore (and as much as twice that offshore), while production costs add around $15/b, and a typical global total (almost evenly split between the two activities) has been about $40/b. Estimates for the largest multinational oil companies (Exxon, BP, RDS, Chevron) show exploration costs ranging from roughly $1.50 to $5.50/b, lifting costs adding $12–18/b and the total costs, with taxes of $3–9/b, at $22–25/b. Conservatively estimated nationwide means for 2016 (inevitably hiding large regional variations) range from $45/b for expensive British oil from the North Sea, to less than $10/b for Saudi and Iranian crude, with the US and Canada averaging about $25/b.

Finally, some representative estimates of the energy cost of oil production. The highest energy return on energy investment (EROEI) came with the discoveries of giant and supergiant Middle Eastern oilfields. Production from some of these huge reservoirs delivered hundreds of times more energy than the cumulative cost of their discovery, development and extraction (including energies for the secondary recovery). Even in the case of two North Sea oilfields, Auk and Forties, the energy invested in their discovery and development was repaid in oil in less than three months. In rich oilfields the net energy ratio can surpass 0.97 or even 0.995. Energy cost of extraction is thus equal to just 0.5–3% of energy in the produced oil, corresponding to EROEI of 33–200. The typical performance of small oilfields is much less impressive. The best estimates of the nationwide US means show that the EROEI for oil discovery and extraction averaged at least 100 during the 1930s, was about twenty-five in the early 1970s and has recently remained between 20 and 30. These findings allow two important generalizations. First, there is no doubt that outside of the Middle East the oil industry has been experiencing a secular decline of its EROEI. However, even the latest, historically low, rates compare favorably with the energy cost of many fossil and non-fossil alternatives.

HISTORY OF OIL PRODUCTION

The historical progress of oil production can be traced fairly accurately from the earliest decades of the industry. The best estimate puts crude oil production (from seeps, pools and hand-dug wells) at about 300t in 1850; by 1880 the total had surpassed 4Mt; by 1900 the total was 22.5Mt (half of it from Russia, and 95% of that from Baku, and 9.5Mt from the US). Between 1900 and 1920 extraction nearly quadrupled to almost 100Mt, and then it doubled in just a decade to 196Mt in 1936, and on the eve of World War II it rose to 272Mt. Another doubling had taken place by 1950 and yet another during the 1950s (to 1.052Gt in 1960), followed by a 2.2-fold increase during the 1960s to 2.35Gt in 1970. The peak was reached at 2.87Gt in 1974; then the production experienced a one-year decline only to reach a new record high of 3.23Gt in 1979. OPEC's second oil-price increase finally made the markets work: by 1983 production had fallen to 2.76Gt (a 15% decline) and it did not surpass the 1979 peak until 1994. By the century's end it stood at 3.61Gt and in 2015 it reached nearly 4.4Gt (see figure 22).

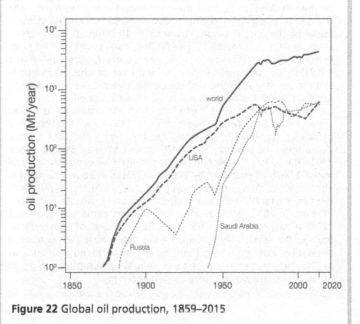

Figure 22 Global oil production, 1859–2015

HISTORY OF OIL PRODUCTION (*cont.*)

There have been also some notable shifts, deletions and new appearances on the list of leading producers. In 1950 the US was by far the largest producer and the USSR was a distant second. Fifteen years later the US was still in the lead, followed by the USSR, Venezuela and Kuwait, which was slightly ahead of Saudi Arabia. Norway did not produce a single barrel and China was extracting just over 200,000bpd. The US lost its primacy to the USSR in 1975 and fell to third place behind Saudi Arabia in 1977. Saudi production peaked in 1980 (at about 85% of the Soviet level) but within five years it declined by two-thirds and the USSR remained the world's largest oil producer until its dissolution in 1991. As the extraction in the USSR's successor states fell, Saudi Arabia became the world's largest producer in 1992 and it was surpassed by the US in 2014.

Although in 2015 Saudi extraction reached a new record of just above 12Mbpd, the US output of 12.7Mbpd was about 6% higher. But because the Saudi crude oils are, on the average, heavier than the US liquids, the ranking in mass terms is reversed, with the Saudis marginally ahead (568.5 vs. 568.2Mt); calling it even might be the best solution. Russia (with 10.98Mbpd) was in third place, and Canada (4.385), China (4.309), Iran (3.920), United Arab Emirates (3.902), Kuwait (3.096), Venezuela (2.628) and Brazil (2.527) made up the remainder of the top ten producers. The US is still the country with the highest cumulative extraction: between 1859 and 2015 it produced nearly 225Gb of crude oil.

The fields that contributed most to this total are the two supergiants, East Texas (flowing since 1931) and Prudhoe Bay in Alaska (producing since 1977), and two California giants, Wilmington (producing since 1932) and Midway-Sunset (since 1901). Russia has the second highest cumulative production, with about 178Gb of crude oil by 2005, and Saudi Arabia is third with about 150Gb. Oil's extraordinarily massive production lends itself to many trivial but revealing calculations and I will cite just three of them. In 2015 global crude oil extraction amounted to almost 92Mb a day, more than the annual oil demand of Switzerland, Israel or Portugal. In 2017 the US and Saudi Arabia each extracted more oil in just two weeks than the entire world did in 1900, and al-Ghawār's output is larger than annual consumption of crude oil in Japan, the world's third largest economy.

Oil transport

The primitive beginnings of oil transport on land included very expensive deliveries by horse-drawn wagons carrying wooden barrels, and short, leaky wooden pipelines (the first one was built in Pennsylvania in 1865). In 1878 the first major cast-iron pipeline (15cm in diameter) linked Bradford and Williamsport in Pennsylvania and a year later it was extended to Bayonne in New Jersey (hence known as the Tidewater line). Expansion of pipelines coincided with the introduction of inexpensive steel, made first by the Bessemer process and later by open-hearth furnaces. Steel, with its high tensile strength, is superior to cast iron, and a fundamental innovation that opened the way for an eventual large-scale expansion of pipeline transport was the invention of the pierce rolling process for the production of seamless steel pipes by Reinhard and Max Mannesmann at their father's file factory in Remscheid in 1885. A few years later they introduced the pilger rolling process that reduces the diameter and wall thickness of pipes while increasing the tube length. The universally used combination of these two techniques is known as the Mannesmann process.

Perhaps the most important non-technical development in the early history of American pipelines was their first government regulation, the Hepburn Act passed by Congress in 1906: by making all interstate pipelines common carriers it guaranteed service to all customers at equal cost. Between the two world wars the US remained the only major economy with an increasingly dense pipeline network. During the 1930s came the first lines carrying refined products and these were greatly extended during World War II as new large-diameter lines carried crude and products from Texas and Oklahoma to the largest consuming nodes in the Northeast.

About two-thirds of all US liquid fuel shipments are now carried by pipelines, and because of the transport of refined products the total volume carried is more than twice as large as the country's total crude oil consumption. The Texas–Louisiana coast, Cushing in Oklahoma, Chicago, New York and Los Angeles are the nation's key hubs for oil and product shipments. In 2013 the country had about 307,000km of liquid fuel pipelines, and about 32% of them were crude oil lines radiating mainly from Texas, Oklahoma and Louisiana. The most notable US pipeline is the Trans Alaska Pipeline System (TAPS) built between 1975 and 1977 to carry oil (original capacity of 1.2Mbpd) from the supergiant Prudhoe Bay field to the ice-free port of Valdez on Alaska's southern coast. Although it broke no length or diameter records (1,280km, 120cm), virtually the entire route crosses permafrost terrain and the pipe had to be built on elevated supports above ground and heated to 60°C. Expansion of American oil production has resulted in continued extension of the country's pipeline grid: in the ten years between 2003 and 2013 the total length of crude oil pipelines increased by nearly 25%, and more lines are being added to accommodate the continuing expansion of shale oil extraction.

Russia's record of pipeline construction is as old and as notable as the US achievements. The first Russian pipeline from the Baku fields to Nobel's refinery was completed in 1878 and in 1896 the Russians began to build one of the first long-distance product pipelines to carry kerosene from Baku (on the Caspian Sea) to Batumi on the Black Sea; the 835km line was completed in 1906. But a large-scale expansion of major trunk lines came only after World War II as an inevitable consequence of the country's enormous territory and of the concentration of its new oil discoveries far from the main centers of consumption in the European part of the country: by 1950 the USSR had about 5,400 km of oil pipelines, and by the time of the country's demise their length had reached 94,000 km. Because the territory of the former USSR was more than twice as large as that of the US, the US had retained its primacy as the country with the world's densest network of pipelines but the USSR surpassed it in terms of the longest large-diameter lines.

Construction of the 3,662km-long Trans-Siberian line from Tuymazy to Irkutsk began in 1957 (completed in 1964), and in 1959 it was decided to build a branching Druzhba trunk pipeline to supply East Germany (via Poland) through the northern spur, and Hungary and Czechoslovakia through the southern branch, with a total length of more than 6,000km and mostly with a pipe diameter of 102cm. Later additions extended its overall length to about 8,000km. In 1973 construction began on the Ust'-Balik-Kurgan-Almetievsk line, 2,120km long and with a diameter of up to 122cm, to carry up to 90Mt of crude oil annually from the supergiant Western Siberian Samotlor oilfield to European Russia, where it connects to the older lines that now take Western Siberian oil all the way to Central and Western European markets.

Europe's rapid post-1960 conversion from coal to imported crude oil led to the construction of many new pipelines, and after major oil discoveries in the North Sea new undersea lines brought fuel to the UK, Norway and Denmark. Rising exports led to large-scale pipeline projects in a number of major producing countries, above all in Saudi Arabia and in Iran, and new discoveries of giant oilfields required the construction of the first long-distance pipelines in China. A pipeline from Kazakhstan (Atyrau) to China's Xinjiang (total length of 2,229km and annual capacity of 20Mt) was completed in 2009, and the Eastern Siberia–Pacific Ocean pipeline – a 4,857km-long link from Taishet (Irkutsk *oblast*) to Kozmino port, near Nakhodka on the Pacific – was built (including a spur to China) between 2006 and 2012.

ADVANTAGES OF PIPELINE TRANSPORT

The preference for pipelines as principal long-distance carriers of oil is above all a matter of logistics and cost. Replacing a 1,000km pipeline carrying 20,000t of oil a day by tanker trucks (assuming each truck holds 25t and covers 1,000km a day) would need a fleet of 1,600 vehicles with a load arriving every 54 seconds. No land transport can be cheaper and only large-scale water-borne transport, that is, large river barges and ocean tankers, can move oil less expensively. But pipelines operate with unmatched reliability and safety and hence with minimal environmental impact. They are also quite compact (1m-diameter pipeline can carry 50Mt of crude oil a year) and are made from a relatively inexpensive material (steel). Common outside diameters range from 60 to140cm, and pipe sections are usually 18–22m long.

Oil is pushed through pipelines by centrifugal pumps powered by electric motors, diesel engines or gas turbines that are located at the origin of a line and then at intervals of 30–160km (depending on the terrain crossed by the line and its throughput). Typical oil speed is 5–12km/hour (hence, it may take up to three weeks for Texas oil from Houston to reach New York) and pipelines normally operate non-stop throughout the year, with sections shut down briefly for scheduled maintenance. Crude oil, different refined products or their different grades (especially those of gasoline) are moved through pipelines according to advance schedules. When two very different batches (such as gasoline and diesel) mix at a transport interface they must be reprocessed on arrival. Batches can also be separated by pigs – polyurethane plugs pushed by the transported liquid. Other pigs (with abrasive coatings) are used to strip deposits from the interior of pipes, and smart pigs, introduced during the 1960s, storing data on onboard computers or relaying it by telemetry, are used to inspect the integrity of pipes and detect any leakages.

Pipelines have the lowest cost for transporting crude oil and oil products on land: while it costs only $3.80–4.50 to move a barrel of Canadian crude from Alberta to Texas, shipping it by rail costs $11–17. The energy cost of oil transportation is also relatively low, both in absolute terms and (even more so) when compared to the aggregate energy that they deliver during the course of their long operational life. Steel, the dominant material in pipeline construction, is produced with an equivalent of about 85t of oil per kilometer for a 60cm-diameter pipeline whose construction requires an equivalent of about 35t of oil per kilometer for a total cost of about 120t of oil, equivalent, or less than 0.1% of the energy in the oil that the pipeline will carry during (at least) four decades of service.

Early waterborne oil transport was as primitive as the first land transfers: the brig *Elizabeth Watts* was the first ship to take such a shipment (about 200t of oil stored onboard in kegs) from Philadelphia to London in 1861. Ships with small (and often leaky) built-in iron tanks followed during the 1860s and 1870s, but the first true tanker (carrying oil against the hull in eight compartments, with the gross tonnage of 2,300t) was the Newcastle-built German *Glückauf* launched in 1884 (it ran aground near Long Island in 1893). Subsequent growth of both typical crude oil tanker capacities and ship sizes was slow. The limited size of the pre-WWI oil market (ships and trains were coal-fueled, automobile ownership had begun to expand only in the US, and there was no commercial flying) did not require large oil tankers.

The largest vessels reached more than 20,000dwt by 1921 and during the 1920s came better interior framing, to make tankers lighter yet sturdier, and better pumps and pipes, but the typical sizes remained small: during the late 1930s tankers rarely carried more than 10,000dwt. Rapid growth of tanker capacities began only after World War II when the mass-produced T-2 tankers (16,000dwt) were not needed by the military after the conflict ended and when some of the even more numerous Liberty class ships (also 16,000dwt) were converted to tankers: the 1946 purchase of 100 Liberties by Greek ship owners became the foundation of the country's subsequent dominance of this shipping sector. Rapidly rising post-WWII demand for crude oil in Europe and Japan then stimulated the development of larger tankers and sizes began to double in less than ten years. As a result, the supertanker designation shifted from 50,000dwt ships (before the mid-1950s) to 100,000 and then to 200,000dwt vessels.

Suezmax ships (up to 200,000dwt and, after the 2009 deepening, maximum draft of 20.1m) can pass through the canal and shorten the trip from the Middle East to Europe, particularly when they are loaded at Yanbu, on the Saudi Red Sea coast. But the additional distance around Africa, or the long transport

WORLD'S LARGEST TANKERS

In 1959, the *Universe Apollo* was the first 100,000dwt ship; then, in 1966, Ishikawajima-Harima Heavy Industries launched the 150,000dwt *Tokyo Maru* and later in the same year the 210,000dwt *Idemitsu Maru*. By 1973 there were more than 350 very large or ultra large crude oil carriers (with capacities in excess of 300,000dwt), and ships of 1 million dwt were expected to arrive soon. They were never launched as tanker sizes peaked with the order of *Seawise Giant* in 1975 and its enlargement three years later. The world's largest ship was hit in 1988 during the Iran–Iraq war but it was subsequently repaired and the 564,763dwt (and nearly 459m long) vessel was relaunched and between 1991 and 2004 operated as the *Jahre Viking*. Renamed once more (*Knock Nevis*), it was moored at Qatar's al-Shahīn oilfield where it served as a floating storage and offloading unit until 2009, when it was sold to Indian ship-breakers and renamed *Mont* for its last journey to Alang in Gujarat where tankers are run aground on a beach and laboriously dismantled.

Figure 23 *Iwatesan*, a 300,000dwt double-hulled VLCC built by Mitsui Engineering & Shipbuilding in 2003

WORLD'S LARGEST TANKERS (*cont.*)

Supertankers stopped growing not because of insurmountable technical problems or because of excessive cost (economies of scale decline rapidly with size but they still continue), but rather because of operational considerations (see figure 23). Very large crude carriers (VLCC, between 240,000 and 350,000dwt) and ultra large crude carriers (ULCC, 350,000–500,000dwt) can dock only at a limited number of deepwater ports or must offload their cargo at special offshore terminals connected to onshore storage sites and refineries by pipelines (and before these were in place they had to offload their cargo to smaller vessels). A limited number of ports of call makes the use of ultra large tankers less flexible, and because of their deep draft supertankers must also follow restricted routes in near-shore waters or channels and they require very long distances while maneuvering and stopping. For example, a typical 250,000dwt tanker (more than 350m long, 60m wide, with a draft of 25m and average speed of about 12 knots) takes over 3km (or fourteen minutes) to come to a full stop, and its turning diameter is 1.8km.

routes between the Persian Gulf and Japan, South Korea and China, make little difference to the overall cost of tanker shipping. The vessels are propelled by cheap diesel fuel; incremental costs are thus quite negligible and oil transport by supertankers adds only about half a cent to the retail cost of a liter of gasoline. Some tankers are equipped with steam heating coils in their cargo compartments in order to keep very heavy or highly waxy crude oils or heavy refined products above their pour point.

The energy costs of tanker shipping are equally marginal. Moving Alaskan oil 3,800km by tanker from Valdez to Long Beach in California requires energy equivalent to only about 0.5% of the transported fuel. And a 300,000dwt supertanker needs an equivalent of only about 1% of the fuel it carries in order to travel more than 15,000km from Ra's Tanūra, the world's largest loading oil terminal on the Saudi coast of the Persian Gulf, to the US East Coast. Declining US oil imports have made this run much less common (even at the time of its peak imports the largest US suppliers were the country's two neighbors, Canada and Mexico, and Venezuela, followed by Africa). But the routes from the Persian Gulf to Europe (via Suez or around the Cape of Good Hope, a trip of forty days) are as busy as ever and Persian Gulf–East Asia shipments have multiplied since China switched from a net exporter of crude oil (last time in 1994) to the world's third largest importer (surpassing Japan by 2009), behind the EU and US.

The route to East Asia leads through the world's two most notorious shipping chokepoints, the Strait of Hormuz, between Iran and Oman, and the Strait of Malacca, between Malaysia and Singapore. More than a third of global oil exports now pass through the Strait of Hormuz, whose shipping channel is just over 3km wide at its narrowest point. Most of the shipments for the world's third (China), fifth (Japan) and sixth (South Korea) largest crude oil importers pass through the Strait of Malacca, whose narrowest passage, the Phillip Channel in the Strait of Singapore, is less than 3km wide. Other notable chokepoints with high potential for collisions, sabotage or terrorist attacks are the Turkish Straits (the Bosporus and Dardanelles), which have to be passed by all tankers taking oil from the Black Sea to the Mediterranean; Bāb al-Mandāb, between Djibouti and Yemen, linking the Arab Sea and the Red Sea; the Kattegat and Skagerrak, between Denmark and Sweden (to pass from the Baltic to the Atlantic); and, of course, the Suez and Panama Canals (nearly 5Mbpd go through Suez and the SUMED pipeline, but less than 1Mbpd through Panama).

Finally, a few comparisons to show the global oil (and refined products) trade that was made possible by tankers and pipelines in several revealing perspectives. In 2015 about 45% of all produced crude oil (close to 2Gt) was sold abroad and the OPEC countries accounted for nearly 57% of that total, virtually all of it transported by tankers.

With low prices (OPEC's reference basket of representative crude streams was just $50/b), global fuel exports were worth only about $1 trillion in 2016 or less than 5% of the global merchandise trade. This share was higher in the recent past when crude oil trade added up to nearly $1.6 trillion in 2013 and to $1.4 trillion in 2005. In addition, about 1.1Gt of refined oil products were traded in 2016, with the US in the lead (about 15% of the global total), followed by Russia and by three countries with large refining capacities, the Netherlands, Singapore and South Korea. Although nearly 50 countries export crude oil and nearly 150 import it (or buy refined products), the oil trade has always been highly concentrated, both in terms of sales and purchases. The world's largest exporters were Saudi Arabia (about 18% of the global total), Russia (about 13%), Iraq, United Arab Emirates, Canada and Nigeria. The EU was the largest crude oil importer in 2016 (buying about 24% of all traded crude), followed by the US (about 19%), China, India and Japan. The EU was also the largest importer of refined oil products, while the US was an increasing net exporter of them.

Lastly, one fact concerning global crude oil trade is rarely pointed out: it is by far the single most massive traded commodity. In 2016 its exports surpassed 2Gt compared to about 1.4Gt of coal and almost the same total for iron ore, and just over 800Mt of natural gas.

Crude oil refining and its products

Some varieties of crude oil can be burned directly in boilers and less than 1% of the fuel's global output is still used raw to generate electricity in a few oil-producing countries. Refining, a paragon of value-adding treatment, involves four basic categories of physical and chemical processes. Distillation (fractionation) separates individual cuts (fractions, that is, fuels with more homogeneous composition), by heating crude oils first in atmospheric and then in vacuum distillation towers. The yield of light distillates is increased by isomerization, catalytic reforming, alkylation and polymerization. Heavy distillates are subjected to thermal and catalytic cracking. And a variety of processes are used to remove unwanted trace compounds and to prepare environmentally more acceptable final products that are shipped to specific markets. I will explain the essentials of these operations through a combination of a brief historic review of advances in refining and a description of a typical process sequence in a large modern refinery.

All early refineries relied on simple thermal distillation, using heat delivered as high-pressure steam in order to separate crude oils into their principal fractions. Consequently, if a particular crude oil contained only a small proportion of light fractions, its thermal refining produced largely medium and heavy liquids. This mattered little until the 1890s when most of the final demand for refined products consisted of kerosene and lubricating oils, but it was a highly unsatisfactory outcome once growing car ownership led to a rapidly increasing demand for gasoline and soon afterwards for diesel oil. Unfortunately, as already explained, most crude oils are not rich in light fractions. Straight thermal distillation of the medium and heavy oils that dominate the global market would yield only 10–15% of the charged volume as the lightest fraction, and the worldwide extent of driving and flying would be restricted by crude oil quality. The industry needed a process that would break C-C bonds to produce lighter compounds and catalytic cracking provided the solution.

CATALYTIC CRACKING

The first breakthrough in producing lighter products from heavier feedstocks came in 1913 when William M. Burton patented thermal cracking of crude oil. Burton's process relied on the combination of heat and high pressure to break heavier hydrocarbons into lighter fractions. A year later Almer M. McAfee patented the first catalytic cracking process that became commercially available by 1923: crude oil was heated in the presence of aluminum chloride, a compound able to break long-chained hydrocarbon molecules into shorter, more volatile compounds, and gasoline yield was as much as 15% higher compared to thermal cracking. But because the relatively expensive catalyst could not be recovered and reused, thermal cracking (less effective but simpler and cheaper) remained dominant until 1936, when Sun Oil's Pennsylvania refinery in Marcus Hook installed the first catalytic cracking unit designed by Eugène Houdry to produce high-octane gasoline.

Houdry's fixed-bed process allowed for the recovery of the catalyst but it required a temporary shutdown of the refining operation while the aluminosilicate catalytic compound was regenerated. Soon afterwards Warren K. Lewis and Edwin R. Gilliland replaced Houdry's fixed catalyst with a more efficient moving-bed arrangement whereby the catalyst circulated between the reaction and the regeneration vessels. This process boosted gasoline yields by 15% and by 1942 90% of all aviation fuel produced for the US war effort was made using this system. An even higher yield was achieved with the invention of powdered catalyst suspended in the air stream (and behaving like a fluid) by four Standard Oil chemists in 1940.

Fluid catalytic cracking (FCC) takes place in a reactor under high temperature (540°C) in less than four seconds. The process of FCC was further improved, beginning in 1960, with the addition of synthetic zeolites. Zeolites are crystalline aluminosilicates whose uniformly porous structure provides an exceptionally active and stable catalyst, improving the gasoline yield by as much as 15%.

The last fundamental addition to the techniques of oil refining came during the 1950s when the Union Oil Company developed the process of hydrocracking (proprietary name, Unicracking). This process combines catalysis at temperatures above 350°C with hydrogenation at relatively high pressures, typically at 10–17MPa. Large-pore zeolites loaded with a heavy metal (platinum, tungsten

> ### CATALYTIC CRACKING (*cont.*)
>
> or nickel) are used as dual-function (cracking and hydrogenation) catalysts. The main advantage of the process is that high yields of gasoline are accompanied by low yields of the two lightest, and least desirable, alkanes (methane and ethane).

Modern refining consists of a number of complex (sequential and feedback) operations arranged and optimized to convert crude oils into the most valuable combination of specific products (see figure 24). The process starts with desalting that removes not only inorganic salts (whose presence would corrode the refinery pipes, units and heat exchangers) but also suspended solids and water-soluble trace metals. Desalted and dewatered crude oil is heated and led to an atmospheric crude distillation unit (CDU) in whose tower the products are separated by different boiling points into principal cuts that include gases, straight-run gasoline, light and heavy naphtha, kerosene and atmospheric gas oil. Petroleum gases (light ends) include small alkanes (methane to butane) with boiling points below 0°C. The two lightest gases – methane (CH_4, with boiling point at −161°C) and ethane (C_2H_6) – are sent into pipelines or burned to produce energy for refinery operations; processing separates LPG (liquefied petroleum gases, mostly propane and butane) that is used as chemical feedstocks or sold for use as household and industrial fuel for space and process heating and cooking.

Light (straight-run) naphtha (compounds with 5–7 carbons) boil away at 27–93°C and heavy naphtha (6–10 carbons) at 93–177°C. Kerosene is a mixture of alkanes and aromatics with 10–14 carbons and it separates at between 175 and 325°C. Light gas oil (diesel oil), with alkanes containing 14–18 carbons, boils at between 204 and 343°C, and is the heaviest cut from a CDU; besides being the second most important road transport fuel it is also used as heating oil and a chemical feedstock. Streams coming from a CDU are first sent

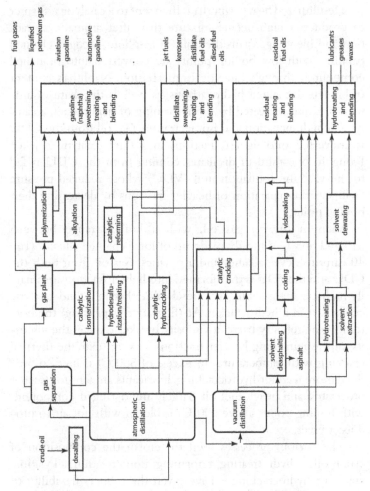

Figure 24 Flow of products in a modern refinery (simplified)

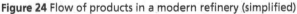

to a hydrotreater where they are heated to the maximum of 430°C, mixed with hydrogen and passed over cobalt or molybdenum catalysts: hydrogen's catalytic reaction removes most of the sulfur and nitrogen present in the liquid as hydrogen sulfide and ammonia.

Desulfurized heavy naphtha is then sent to a catalytic reformer to produce a high-octane mixture that, after aromatic extraction and blending, forms the bulk of gasolines. Straight-chained paraffins with low boiling points are converted into branched compounds through catalytic isomerization. Naphtha's content of branched-chained hydrocarbons, as well as the share of aromatics, is also increased by the reforming of naphthenes, a catalytic dehydrogenation and isomerization. Gas oil from the CDU is converted into lighter fractions in a fluid catalytic cracker. Roughly one-third of distillates coming from the CDU is fed to the vacuum distillation unit (VDU) whose reduced pressure facilitates the separation of heavier fractions by decreasing their boiling point.

Vacuum (heavy) gas oil, with a boiling range between 315 and 565°C is mostly made up of long-chained alkanes (16–40 carbons), cycloalkanes and aromatics. Gas oil from both the CDU and the VDU is transformed into lighter fractions (primarily high-octane gasoline and diesel oil) either in a fluid catalytic cracker or in a hydrocracker. At (figuratively speaking) the bottom of the refinery barrel a viscosity breaker reduces the viscosity of the remaining heavier fractions, coking (extreme thermal cracking with temperatures in excess of 500°C) is used to produce more feed for hydrocracking, lubricants are sent for further processing and heavy fuel oil (largely multi-ringed compounds with boiling points above 600°C) is blown with hot air to produce asphalt.

The various processes used to reform the composition of crude oils – hydrotreating, reforming, isomerization, alkylation, cracking, hydrocracking – have given the refiners flexibility to adjust (within economical limits) the shares of final products. While a typical breakdown of atmospheric distillates from a good quality (light) crude oil would be less than 30% light and heavy naphtha, about 10% kerosene, 15% gas oil and 45% residue, modern refineries can reduce the residue to just 20% and boost the share of gasolines to almost half of the total output. These adjustments make it possible to meet specific national and regional demand. The US, with its large fleet of passenger cars, now consumes almost half of its refined products as motor gasoline (46% in 2015) and about 30% distillate fuel oil (mostly ultra-low sulfur diesel). In contrast, in 2015, gasoline was 28% of Japanese refinery output but heavy fuel oils and diesel oil (used by industries and in shipping) accounted for 35%. Important shifts, reflecting changing demands, also take place over time. Since 1945 gasoline has accounted for roughly the same share of the US refinery output (about 45%) but the percentage of heavy fuel oil fell from about 20% to less than 5%, while that of kerosene rose from less than 1% to nearly 10%. A closer look at the most important refinery products makes clear their unique qualities and specific uses.

With annual output of 1Gt in 2015 gasolines are the world's most important refined products in terms of price; in mass (or volume) terms gas and diesel oils are more important globally (in 2015 about 1.3Gt) as well as in the EU, Japan and China, but not in the US and Canada. Straight-run naphtha, reformed naphtha and streams from hydrocracking, fluid catalytic cracking, isomerization and alkylation go into their blending. Typical gasoline contains roughly 15% straight-chain alkanes (C_4–C_8), 25% branched alkanes (C_5–C_{10}) and the same share of alkyl benzenes (C_6–C_9). All gasolines are highly volatile and flammable, and new environmental regulations require a very low presence (maximum 50ppm) of sulfur. All of them require additives to improve the efficiency of their conversion as well as to prevent oxidation and rust formation. Gasoline was used only very inefficiently in all early internal combustion engines because of a 'knocking' problem that is inherent in Otto's internal combustion engines.

ENGINE KNOCK

In internal combustion engines the compressed fuel-air mixture is ignited from the top of the cylinder by a spark: within half a millisecond an ignition wave starts to propagate downwards, but in the presence of high pressure and high temperature the unignited fuel-air mixture can begin to combust spontaneously, sending a pressure wave in the opposite direction to that of the spreading ignition flame, resulting in a characteristic, and violent, engine knock. Heptanes are particularly prone to this spontaneous ignition caused by compression, while octanes are resistant. Retarding the spark reduces the knocking but the only way to avoid it was to operate the engines at low compression ratios and thus lower their efficiency: in early internal combustion engines the ratio was held below 4.3:1 inevitably limiting the engine efficiency and raising worries about the adequacy of post-WWI crude oil supplies. A General Motors team, led by Charles F. Kettering, researched remedial options and identified ethanol as an effective anti-knocking ingredient that made it possible to use gasolines with a higher share of heptanes.

But large-scale production of ethanol was prohibitively expensive and the only other known alternatives (bromine and iodine) were even more costly. This led the company to search for new inexpensive options. A team led by Thomas Midgley first ran some promising trials with tetraethyl tin before confirming, in 1921, that tetraethyl lead was a highly effective anti-knocking agent even when added at concentrations as low as 1/1000 of the fuel's volume. The first leaded gasoline was marketed in February 1923 and its use allowed the compression ratio of internal combustion engines to climb above five and eventually to reach the modern range of 8:1–10.5:1. Substantial fuel savings in road transportation were not the only benefit: leaded gasolines made it possible to build more powerful (and hence faster) and more reliable aeroengines. But, as already noted, decades of leaded gasoline consumption resulted in dangerous levels of lead contamination in cities. Phase-out of the additive began in the US in 1974, its use was prohibited as of January 1 1996, and most of the world's leading gasoline consumers (including the EU, China and India) stopped using the leaded fuel by the year 2000. By 2016 only Algeria, Iraq and Yemen continued to use leaded gasoline.

Diesel fuel – blended from fractions obtained by atmospheric distillation as well as from cuts produced by hydrocracking, fluid catalytic cracking, viscosity breaking and coking – is usually the second most important product. The fuel is heavier than gasoline (density of 0.83–0.85g/ml, compared to 0.73–0.75g/ml for gasoline) and its lightest category (US No. 1 diesel fuel) is dominated by molecular chains with 9–16 carbons, while the heavier fraction (No. 2) has molecules with 10–20 carbons. Low-sulfur diesel fuel, sold in the US since 1993, still had a relatively high (500ppm) sulfur content but new ultra low sulfur diesel, available since 2006, can have no more than 15ppm S and its use is compatible with catalyst-based emission control devices (diesel particulate filters and NO_x absorbers). The EU has even stricter limits, just 10ppm S since 2009. This less polluting fuel has both lower energy content and lower fuel economy compared to standard diesel. Due to higher compression ratios diesel engines have higher efficiency than gasoline-fueled engines: the difference used to be as much as 30% but it has been shrinking with the introduction of better gasoline engines. In addition to passenger cars, trucks, off-road vehicles and ships, diesel fuel is also used to generate electricity during peak demand hours, but its sale for domestic heating has declined due to the widespread adoption of natural gas.

Kerosene is the second lightest fraction (C_{11}–C_{13}) of crude oil distillation. Much like gasoline, it is a colorless and highly flammable liquid that separates from crude oil at 150–275°C. Although still commonly used for lighting and cooking in many modernizing countries (India is a big consumer) and for seasonal heating both in the US and Europe (mainly in portable kerosene heaters), its most important use is to power jet engines (gas turbines) mounted on the wings or bodies of both commercial and military aircraft. Kerosene is a better aviation fuel than gasoline because its slightly higher specific density (0.81 vs. 0.73g/L) results in an energy density that is about 13% higher than that of gasoline (35.1 vs. 31.0MJ/L), an important consideration when the volume of fuel stored in aeroplane wings and fuselages is obviously limited. Moreover, the slightly heavier fuel is also cheaper than gasoline, and because it is less volatile it has lower evaporation losses at high altitudes and a lower risk of fire during refueling and ground storage, an attribute that also results in less flagrant, and hence more survivable, crash fires.

Fuel oil is a collective label for a large category of distillate and residual liquids with widely differing properties and with a confusing taxonomy. They are usually divided into six classes. Fuel oil no. 1 is essentially the straight-run kerosene, while no. 2 corresponds (except for the additives) to diesel fuel and it is also used as household heating oil. Light fuel oils, distillate fuel oils and gas oil are their common synonyms. No. 4 oil is the most commonly used kind of fuel in powerful marine diesel engines; it contains up to 15% residual oil and hence it is less volatile than automotive diesel (fuel oil no. 2). Fuel oils used in waterborne transportation are usually called bunker fuels. Worldwide statistics show bunker sales of about 170Mt in 2015, with by far the highest national sales in Singapore (more than 40Mt), followed by the US, South Korea and China. The two heaviest fuel oils (nos. 5 and 6) are commonly called residual oils or heavy fuel oils. No. 6 used to be the most common choice to replace coal in electricity generation but due to very high sulfur levels it was replaced by natural gas, or even by coal (with desulfurization). But coal-fired electricity generating plants use heavy fuel oil to initiate the combustion in their boilers (much like kindling for woodfires).

What remains at the bottom of the barrel cannot be used as fuel but it has valuable non-fuel applications. In 2015 the US refineries converted the following shares of crude oil into non-fuel products: 2% into asphalt and road oil, 1% into lubricants and less than 0.1% into waxes; in addition, just over 5% of all crude oil delivered to refineries ended up as petroleum coke. Asphalts are chemically even more complex than residual fuel oils and the unique properties of these thermoplastic (soft when heated, hard when cooled) materials have made them one of the most ubiquitous constituents of our constructed world. Asphalts are either straight-run products or (for higher viscosity) are made by blowing air through hot liquid asphalt. Polymers, including polybutadiene in the form of ground tires, can also be added.

Oil refining is by far the most energy-intensive part of the entire sequence of liquid fuel production. US data shows that, on average, the country's refineries consume an equivalent of about 10% of the energy in the processed crude oil, mostly as hot steam, electricity and gases. After subtracting the non-fuel products (lubricants, tar, asphalt) the net energy content of refined fuels is about 85–88% of the energy in the crude oil. Specific studies indicate that gasoline contains 6–10 times more energy than the energy cost of its refining.

Crude oil refining, producing a wide variety of highly flammable gases and liquids, requires safety precautions in locating the processing and storage facilities, preventing spills and facilitating fire-fighting. Minimum spacing of 60–75m is mandatory for separation of units within a refinery and typical throughputs of about 2.5t of crude oil/m^2 means that a large (500,000bpd) facility requires at least 1,000 hectares, although they can be much larger.

In 2017 the worldwide capacity of more than 650 operating refineries was 97Mbpd. Predictably, the largest consumers of liquid fuels had the largest total refining capacities. The US led with 18.6Mbpd, followed by China (14.2), Russia (6.4), India (4.6) and Japan (3.6). As for the largest oil companies, Exxon Mobil could process more than 5Mbpd, with China's Sinopec, Royal Dutch Shell, BP and Conoco Phillips making up the top five. At the beginning of 2017 worldwide refining capacity was about 20% higher than the actual throughput, implying a utilization rate of 80%. Among the major consumers of refined liquid fuels only Russia and the US were net exporters while the EU, China and Japan were major net importers of refined fuels.

As with so many other industrial processes, the closing decades of the twentieth century have seen a trend toward consolidation and increasing unit sizes (leading to economies of scale) in crude oil refining. For example, in 1980, the US had just over 300 refineries, nearly as many as right after World War II, but by 2000 the total had halved to about 150. Large modern refineries now commonly process in excess of 100,000 barrels per day, that is, at least 5Mt of crude a year, and the top ten produce in excess of 500,000bpd. The world's largest refinery in 2015 was Jamnagar (in India's Gujarat) with capacity of 1.24Mbpd, followed by Venezuela's Paraguana (955,000bpd) and South Korea's Ulsan (840,00bpd). Saudi's Ra's Tanūra (550,000bpd) was the largest refinery in the Middle East and Europe's largest facility was Pernis (404,000bpd) in the Netherlands.

Refineries usually have on-site capacity for storing enough crude oil for two weeks of operation. Voluminous storage is also needed at export oil terminals to fill the tankers and at the starting points of large pipelines to dispatch the custom batches of crude oil or refined products. Much smaller storage tanks are common at many industrial establishments. By far the largest oil storage is the US Strategic Petroleum Reserve that began to fill in 1977 with imported Saudi oil. Crude oil is stored deep underground in four massive salt caverns along the Texas and Louisiana Gulf Coast. The maximum capacity is 713.5Mb and the reserves stood at 685Mb in June 2017, representing about 10% of US annual oil consumption and the equivalent of 149 days of import protection (based on 2015 net crude oil imports), well above the International Energy Agency's requirement of 90 days of import protection. Aggregate oil stocks in OECD countries (including national strategic petroleum reserves, commercial stocks and oil-on-water) were close to 6Gb at the end of 2016, which means that the Western world is now much better prepared to weather any sudden, large-scale oil import interruptions than at any time in the last four decades, and China has followed this example and its strategic petroleum reserve has a target storage of 500Mb.

5
How long will oil last?

This chapter's title is a deliberately loaded question with multiple answers. The most obvious answer – if the question is taken literally and purely in terms of physical presence – is as long as the planet Earth. Even the best conceivable enhanced recovery techniques will still leave behind a substantial amount of the oil originally present in reservoir rocks. More importantly, the excessive cost of discovering and producing liquid oil that is stored in countless small, marginal or virtually inaccessible formations, and unappealing returns on extracting liquid hydrocarbons from most of the known oil sands, oil shales and tar deposits, will ensure that a significant amount of the oil originally in place in the Earth's crust will never be brought to the surface.

If the question means simply how long there will be some commercial production of crude oil then the answer is also easy: definitely throughout the entire twenty-first century. Even those who have argued that the peak of global oil extraction is imminent must concede that we have yet to discover many conventional oilfields and that the new ways of extracting oil from nonconventional sources mean that some of these accumulations will remain in commercial production during the closing decades of the twenty-first century. And while accelerated decarbonization can displace a number of specific fossil fuel uses with non-fossil alternatives, other oil uses will prove much more resistant to change. But I will not try to answer the question of how long crude oil will remain the single most important fossil fuel in the global primary energy supply; nor will I predict when global oil extraction will reach its peak and begin to decline. Answers to these questions are contingent on the unknown magnitudes and trends of many variables, and any new predictions of the peak production year would only extend an already long list of failed forecasts of this kind.

What I will do instead is to provide a proper historical perspective on oil's role in the global energy supply. Then I will survey the fallacies and facts concerning the currently fashionable catastrophic prognoses of an imminent end of the oil era – the sentiment embodied in publications whose titles claim that we have already reached the production peak, that the party's over, or, in the most extreme fashion, in Richard C. Duncan's Olduvai theory, that the decline of oil extraction will plunge humanity back to a life comparable to that experienced by the first primitive hominins who inhabited the famous Tanzanian gorge some two million years ago. Finally, I will look beyond oil, outlining briefly some major means of providing liquid fuels from sources other than conventional crude oil.

Oil in the global energy supply

If oil's importance were to be judged by the frequency with which the words oil, crude oil or petroleum are mentioned in the media or by politicians, the inevitable conclusion would be that no other source of energy is more important for the survival of civilization. This would be an incorrect and indefensibly exaggerated notion. Undoubtedly, liquid fuels have had an enormous impact on the modern way, and quality, of life, but outside of North America they became very important only during the last two generations, after 1960 in Europe and Japan, and since the

1980s in the populous countries of Asia. Global dependence on oil is thus a relatively new phenomenon, and this reality should forcefully remind us that we should not exaggerate the fuel's indispensability: we had ingenious industrialized societies capable of delivering a decent quality of life long before oil consumption rose to its current level – and there will be prosperous societies supporting a good quality of life long after liquid hydrocarbons have become minor constituents of the global energy supply. A telling indication of this transition process is that in 2017, a Google search returns nearly as many hits for natural gas as for petroleum and crude oil combined!

ENERGY TRANSITIONS

Pre-hydrocarbon industrializing societies were energized by coal and also by the generation of hydroelectricity. Coal, the quintessential fuel of the nineteenth century's industrialization, continued to dominate the global supply of commercial energy during the first half of the twentieth century: its share declined slowly, from about 95% of the modern energy supply (excluding traditional biofuels) in 1900 to about 80% of the total in 1930 and to just over 60% by 1950. Meanwhile, crude oil provided just 4% of the world's commercial primary energy in 1900, 16% by 1930 and 27% by 1950 when natural gas supplied about 10% of the total (see figure 25). In many leading economies coal was much more dominant: even by 1960 its share in primary energy supply was nearly 60% in Japan, 61% in France, 77% in the UK and 80% in Germany, at that time fueling the country's impressive economic expansion.

Global transition from coal to hydrocarbons accelerated during the 1960s: 1962 was the first year when coal provided less than half of the world's primary energy. Later, because of oil's numerous advantages, even the two rounds of OPEC's oil price increases could not stimulate coal's comeback. The most important factors that explained coal's continuing retreat were the relatively high cost of its underground extraction, its inflexibility of use and the considerable environmental impact of its production and combustion (the latter including acid precipitation and high carbon emissions). But there is no doubt that the post-1972 OPEC-driven oil price rises

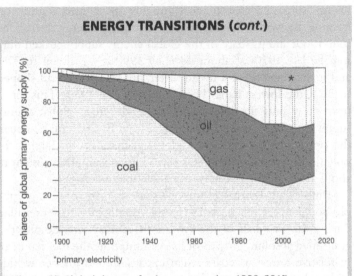

ENERGY TRANSITIONS (cont.)

*primary electricity

Figure 25 Global shares of primary energies, 1900–2015

have definitely helped to slow down coal's global retreat, from about 33% of the total supply of primary commercial energy in 1970 to 28% by the year 2000. But resurgent China disregarded all these drawbacks: it needed enormous amounts of energy fast in order to fuel its rapid rates of economic growth. As a result (and, secondarily, thanks to India's rising demand), coal's share in global primary energy supply rose to 33% during the first decade of the twenty-first century and it was almost that high in 2015.

But outside China and India coal is now in retreat as its most important use, for generation of electricity, is being replaced by cleaner and more efficiently converted natural gas (particularly by combined cycle gas turbines). Most notably, the US dependence on coal-generated electricity declined from 45% in 2010 to 30% in 2016.

The best global aggregate count indicates that during the twentieth century coal and crude oil supplied roughly the same amount of primary energy, each approximately an equivalent of 125Gt of oil, but during the century's second half crude oil's

energy surpassed that of coal roughly by a third. Crude oil became the world's leading source of primary energy during the mid-1960s and its share of the global energy supply peaked during the late 1970s at about 44%; by 1990 it was down to 37% and it remained there a decade later, but by 2010 it had slipped to 32%, the lowest it has been since the late 1950s. This long-term view puts the importance of crude oil in global energy supply into a proper historical perspective. By 2015 oil had been the largest component of the global primary energy supply for only fifty years and during that period its consumption rose from about 1.6 to 4.3Gt a year, nearly a 2.7-fold increase.

But in global terms, crude oil's share of primary energy supply could never reach coal's massive dominance that prevailed during the first half of the twentieth century and, despite the continued absolute growth of its consumption, the fuel has been in relative retreat as coal's Asian expansion and greater world-wide reliance on natural gas combined to supply nearly 60% of all commercial energy in 2016. Oil's relative retreat is also illustrated by the declining oil intensity (units of oil needed to produce a unit of economic product) of all major economies as well as on the global level. When expressing the US GDP in constant ($2011) monies, oil intensity of the US economy fell from about 140kg/$1,000 in 1964 to 55kg/$1,000, a 60% decline in fifty years. During the same period the global decline of average global oil intensity was even slightly greater, amounting to about 66%.

These impressive declines have been achieved through a combination of the higher efficiencies with which we now convert refined oil products and their replacement by gaseous hydrocarbons and non-fossil energies. For example, in 1975, tens of millions of American households (mostly in the Northeast) used fuel oil for relatively inefficient (about 50%) home heating; its replacement by natural gas not only eliminated a previously large market segment but it also lowered the atmospheric carbon burden: even so-called mid-efficiency natural gas furnaces had efficiencies around 70% during the 1980s and well above 80% a generation later, and today's best high-efficiency furnaces convert 97–98% of the chemical energy in methane into heat.

And with the retreat of liquid fuels from electricity generation (less than 7% of all refined fuels were burned in power plants in 2015) and residential uses such as heating and cooking (now also less than 7% of global demand for liquids), the consumption of refined oil products has become even more concentrated in the transportation sector: all major forms of moving goods and people – be it shipping, railroads, trucking, automobiles and flying – rely overwhelmingly on refined fuels. But because this sector has become such a critical component of all affluent economies and lifestyles, it is no exaggeration to conclude that at the beginning of the twenty-first century modern civilization is defined in many important ways by its use of liquid fuels and hence it will go to great lengths to ensure their continued supply.

And it must be repeated that this importance goes beyond the reliance on high-performance fuels in all forms of transportation: oil-derived lubricants are indispensable for countless industrial tasks; modern transportation infrastructures are unthinkable without oil-derived paving materials; and syntheses of scores of plastics begin with oil-derived feedstocks. All of these benefits derive from the extraction of a resource that is not renewable on a civilizational timescale, and since the mid-1990s the questions about its durability have been receiving increasingly worrisome answers from some oil geologists whose arguments have been given prominent coverage by the media. Nearly all of those who have argued that the peak of global oil production is imminent have not foreseen any subsequent comfortable plateau but a rapid decline, and have repeatedly assured us that its inevitable consequence will be the end of modern prosperity and an intensifying fight over the dwindling oil resources. I will deconstruct these scares and show why such bleak scenarios are not likely to prevail.

Oil peaks

The basic assumptions tirelessly repeated by the proponents of the theory of an imminent peak of global oil production are as follows. When we compare the total volume of the estimated ultimate recovery (EUR) of oil with the worldwide cumulative production we see that (depending on the somewhat uncertain size of EUR) we have already extracted half of the EUR or are about to do so in a matter of years. Once the global oil production reaches its peak its fairly steep descent will start fairly promptly because the complete extraction cycle must follow a function described by a normal (bell-shaped, Gaussian) curve (with the area beneath the curve equal to EUR). Given the importance of oil for modern civilization this inevitable decrease in annual oil production will have enormous consequences, and some of the leading peak oil theorists went as far as writing obituaries of modern civilization.

L. F. Ivanhoe, an American geologist, believed that an early end of the oil era will bring 'the inevitable doomsday' followed by an 'economic implosion' that will make 'many of the world's developed societies look more like today's Russia than the US.' Richard C. Duncan, originally an electrical engineer, saw the peak as 'a turning point in human history' leading to massive unemployment, breadlines, homelessness and a catastrophic end of industrial civilization. Indeed, his 'Olduvai theory' had humanity returning soon to the condition in which our hominin ancestors lived nearly two million years ago. But all of the leading proponents of the theory of the imminent peak of global oil extraction (Colin Campbell, Jean Laherrère, L. F. Ivanhoe, Kenneth Deffeyes) have resorted to more or less alarmist arguments. Their writings and speeches have mixed incontestable facts and sensible arguments with indefensible assumptions and caricatures of complex processes as they ignore those realities that do not fit their preconceived conclusions.

Their conclusions are based on simplistic interpretations. Values of EUR are not at all certain, and tend to rise with better understanding of petroleum geology, with frontier exploration and with enhanced recovery techniques. Moreover, the proponents of an imminent peak of global oil extraction disregard the role of prices, they ignore historical perspectives, and they presuppose the end of human inventiveness and adaptability. But it is precisely their bias and their catastrophic message that have attracted the mass media (ever eager to spread new bad news) and impressed a scientifically illiterate public. My critique rests on three fundamental realities. First, these recent peak oil worries are only the latest instalment in a long history of failed peak forecasts. Second, the claim of peak oil advocates that this time the circumstances are really different and hence their forecasts will not fail mixes correct observations with untenable assumptions. Third, and perhaps most importantly, when contemplating a world with little or no oil, a gradual decline of global oil production does not have to translate into any economic and social catastrophes.

Public concerns about running out of fossil fuel resources date to 1865 when William Stanley Jevons, a leading economist of the Victorian era, published a book in which he concluded that it is 'of course . . . useless to think of substituting any other kind of fuel for coal' and that the falling coal output must spell the end of Britain's national greatness. In reality, the UK's coal extraction continued to expand until the second decade of the twentieth century and its subsequent decline had nothing to do with the exhaustion of resources, and everything to do with the arrival of new fuels. In 2015, when the last remaining British underground coal mine was shut down, the UK still had more than 200Mt of bituminous coal reserves, but extracting the fuel had already become an unappealing proposition decades ago when compared to producing crude oil and natural gas from the North Sea, or to importing much cheaper foreign coal. In 2017 the world's proved reserves of coal were about 1.1Tt (implying R/P ratio of more than 150 years) and that total could certainly be multiplied with more exploration and with advanced mining techniques. Clearly, post-1950 switching from coal to hydrocarbons (and to primary electricity) has had little to do with 'running out' of actual mass stored in the Earth's crust.

But with oil the argument has been different, as the proponents of imminent peak production anticipated major declines in the fuel's availability.

OIL'S REPEATED END

Published reports about the imminent end of oil production can be traced as far back as the 1870s, and fears about running out of liquid oil were quite strong in the US during the early 1920s. But the most influential argument was made by M. King Hubbert, an American geologist, who postulated that mineral resource extraction follows an exhaustion curve that has the shape of normal (symmetrical, bell-shaped) distribution: its peak is immediately followed by a decline whose course mirrors the production rise. Hubbert used this approach to accurately predict the peak of US oil extraction in 1970, and the symmetrical exhaustion curve thus acquired the status of an infallible forecasting tool: once the recoverable resources are known and the past production is plotted then a symmetrical continuation of the curve shows the peak extraction year, declining production and the timing of eventual resource exhaustion.

Hubbert's own forecast put the peak of global oil extraction between the years 1993 and 2000. In reality, global output in 2017 was nearly 40% above the 1993 level, a substantial error that clearly invalidates the original forecast. Hubbert's fame rested more on his supposedly accurate forecast of US peak oil production in 1970. Indeed, the output peaked at 11.3Mbpd in 1970 (and it was down by 10% ten years later, conforming to predicted

OIL'S REPEATED END (cont.)

mirror decline) – but Hubbert's forecast pegged it 20% lower and based it on EUR of 200Gb, the total that he himself had raised from the 150Gb that he had estimated just a few years earlier. But between 1859 and 2005, before hydraulic fracturing began to make a substantial difference, the US oil industry had already produced 192Gb, it was still the world's third largest producer of crude oil, and had 32Gb of remaining reserves. And the difference between Hubbert's forecast and reality has grown much wider with the post-2005 rise of shale oil extraction.

As a result, the post-peak decline of US oil production has not been a mirror image of the incline, and the country's oil extraction has not followed a symmetrical curve: in fact, in 2017, the US crude oil output was on the way to surpassing the record set in 1970, and to reach a level nearly five times higher than Hubbert's forecast for 2017 (see figure 26). This could hardly be seen as an admirable forecasting record. Hubbert underestimated the conventional recovery because he had no knowledge of the Prudhoe Bay supergiant oilfield or of coming giant finds in the Gulf of Mexico, and (as everybody else) he did not even consider that nonconventional resources could become major sources of new supply. Obviously, any EUR is just that, an estimate subject to revision, not a fixed total.

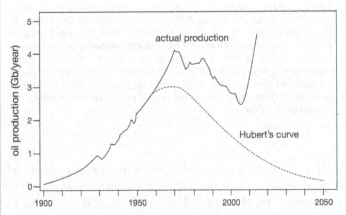

Figure 26 Hubbert's forecast of US peak oil production and the subsequent extraction decline – and the actual 1970–2016 course

OIL'S REPEATED END (cont.)

Hubbert has had many followers who have been oblivious to the fact that actual production has not been conforming to his forecasts. In 1977 the Workshop on Alternative Energy Strategies set the global oil peak as early as 1990 and most likely between 1994 and 1997. In 1978 Andrew Flower wrote in *Scientific American* that 'the supply of oil will fail to meet increasing demand before the year 2000'. In 1979 the CIA believed that the global output must fall within a decade. In 1990 a USGS study put the peak of non-OPEC oil production at just short of 40Mbpd before 1995 – but the actual output was more than 47Mbpd in 2005. Some peak-of-oil proponents had already seen their forecasts fail: Campbell's first peak was to be in 1989, Ivanhoe's peak was in 2000, Deffeyes set it first in 2003 and then, with ridiculous accuracy, on Thanksgiving, 2005.

One would think that this record would dissuade more entries, but the true believers could not resist enlarging this list of failures. In the early years of the twenty-first century we were told that the failure of forecasts produced during the 1970s or 1980s should not be used to argue for a high probability (or certainty) of future failures. Their principal post-2000 argument was that, unlike 20–30 years earlier, exploratory drilling had already discovered some 95% of the oil that was originally in place in the Earth's crust. Consequently, more frantic and more extensive drilling efforts would make no difference: all it would do is discover the small volume of remaining oil faster. And, the peak oil proponents also argued that efficiency improvements (even outlawing SUVs) or a new-found frugality in affluent countries cannot make any fundamental difference: the slower increase in global oil demand (the increase itself being guaranteed by the huge oil needs of modernizing countries) would only slightly postpone the timing of the peak.

To sum up, the entire notion of an imminent peak of global oil production is based on three key claims: that recoverable oil resources are known with a high level of confidence; that they are fixed; and that the history of their recovery is subsumed by a symmetrical production curve. None of these claims is true.

EUR cannot be known with a high degree of confidence and it is not fixed as long as large areas of the Earth remained unexplored or only cursorily assessed, and as long as technical advances are able to transform previously unexploitable resources into major sources of new supply. But even before the re-evaluation of global oil reserves (due to the addition of nonconventional deposits whose recovery became possible thanks to technical advance), and before the rapid ascent of America's oil extraction using hydraulic fracturing, the claims about an imminent peak of global oil extraction followed by a precipitous demise of the oil era ignored several fundamental facts.

To begin with, if the peak of oil extraction is coming soon, should not then the prospective shortage of the precious fuel result in relentlessly rising prices and should not buying a barrel of oil and holding on to it be an unbeatable investment? These conclusions are patently wrong. Oil is not getting either intolerably more expensive to find or to develop, and a barrel of crude, say West Texas intermediate, bought at $17.65/b at the end of 1986 (that is, $39.38 in 2017 monies) and sold in June 2017 at $50/b would have earned (even when assuming no storage costs) an average annual return of 1.1%, a performance vastly inferior to any guaranteed investment certificates and truly a miserable gain when compared with virtually any balanced stock market fund (and a 1980 barrel sold in 2017 would have resulted in a 50% loss!).

Then there is the basic property of a bell-shaped curve. Even if a peak oil promoter had believed in 2005 that the global oil extraction would start declining in a matter of years, the Hubbertian reality would demand that half of all oil was yet to be extracted after that date, and the decline close to a normal curve shape would mean that we would still have more than a century of oil production ahead of us. This means, for example, that a symmetrical curve with the peak at 83Mbpd in 2010 would indicate global extraction of roughly 65Mbpd in 2030 and nearly 50Mbpd by 2050. Clearly, the oil era would not be nearly over even if we had already reached the peak in oil production in the recent past.

And, quite inexplicably, those who forecast an imminent peak of global oil production and a rapid end of the oil era completely ignore fundamental, and proven, economic realities and assume that future demand is immune to any external factors. This is patently false: an indisputable peak followed by precipitous decline in production would not trigger an unchecked bidding for the remaining oil but would rather accelerate an ongoing shift to other energy sources. OPEC learned this lesson in the early 1980s when record high prices were followed not only by the decline of oil's share in global energy supply but also by an abso-lute decline in global oil demand and a drastic fall in price (by about 60% for an average OPEC barrel between 1981 and 1986).

Consequently, even if resources were rather constrained and we found ourselves on the declining right-hand side of the pro-duction curve, we would still have many decades of oil era left and oil prices would not reach an astronomical level. In real-ity, we are still on a gently ascending slope and we also have plenty of evidence that the world has more undiscovered oil than the most pessimistic past estimates would indicate. Most notably, the most comprehensive assessment of the world's undiscovered resources of conventional oil, published by the US Geological Survey in 2000, offered the following division among the three key categories: by the late 1990s roughly 710Gb of oil had been produced worldwide, leaving about 890Gb of remaining known reserves; nearly 690Gb of oil were to come in the future from reserve additions in currently known fields, and roughly 730Gb were yet to be discovered, giving an EUR of about 3.020Tb. In 2012 the USGS updated its worldwide estimate and put the total of undiscovered conventional resources that would be techni-cally recoverable at 565.3Gb of crude oil and 166.7Gb of natural gas liquids, or a total of 731Gb. And even the 95% confidence limit (near certainty) of the original estimate of the undiscovered reserves of conventional oil was 400Gb, that is, nearly three times as much as a typical claim by those who saw an imminent peak of global oil extraction.

According to the USGS about 60% of all undiscovered reserves are almost equally split among three regions, Latin America and the Caribbean, sub-Saharan Africa, and the Middle East, with the following six basins having the largest discovery potential: the Mesopotamian Foredeep Basin, the West Siberian Basin, the as yet completely unexplored East Greenland Rift Basin, the Zagros Fold Belt, the Niger Delta and the Rub' al-Khālī Basin of eastern Saudi Arabia. In North America, the best prospects for major new oil discoveries are in northern Alaska, in the Canadian Arctic and in the Gulf of Mexico. In Latin America, large reserve additions will come in Venezuela and in Brazil's offshore waters, perhaps most importantly in Foz do Amazonas, in the delta of the river. Most of Africa's untapped oil resources are in waters off the Congo and Niger, but significant potential remains in Algeria and Libya. In the Middle East, both of the two leading producers, Saudi Arabia and Iran, will see substantial new discoveries, as will Iraq.

Given all of these uncertainties, a large number of future production curves can be drawn on the basis of different estimates of ultimate conventional oil recovery (see figure 27). For example, the EUR of about 3Tb would (depending on the future rates of consumption) imply a peak of conventional oil extraction sometime after 2020 and it would mean that global production during the 2040s could still be as high as it was in the early 1980s. But the noted assessments offer an inappropriately narrow resource perspective as they account only for conventional resources. As Jean Laherrère (one of the most vocal proponents of an early peak oil production) conceded,

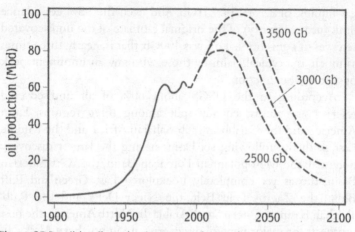

Figure 27 Possible oil production curves during the twenty-first century

with the addition of the median reserve estimates of natural gas liquids (200Gb) and nonconventional oil (700Gb) there would still be some 1.9Tb of oil to be produced, double the amount of his estimate for liquid crude oil alone. Since that time, we have gained a better understanding of nonconventional resources, and the success of US shale oil extraction proved how technical advances can convert a significant share of such resources into economic reserves.

At the beginning of 2010, BP's annual survey put the world's conventional crude oil reserves at 1.476Tb after adding reserves in Canada's oil sands, in Venezuela's Orinoco heavy oil belt and in the US oil shales. The total stood at 1.707Tb in mid-2017, more than a 15% gain in just seven years. Extraction of North America's nonconventional oil is already changing the shape of both US and Canadian EUR curves, and more reserves will be added in the future, even in the regions that were previously explored for conventional resources. The latest example of such a gain is the USGS announcement in November 2016 that credits Wolfcamp shale in West Texas with 20Gb of crude oil recoverable with current practices (and 16 Tcf natural gas): at almost three times the total in North Dakota's Bakken shale this would make it the country's largest crude oil deposit. Many other countries have extensive shale deposits, and in 2015 the US Energy Information Administration put the worldwide total in unproved technically recoverable reserves of shale (tight) oil at nearly 420Gb: outside North America the largest potential is in Russia, Argentina, China, the United Arab Emirates, Libya and Kazakhstan.

And the coming shape of exhaustion curves will also be greatly influenced by adjustments in demand, a phenomenon clearly demonstrated by the decline and stagnation of global oil consumption between 1979 and 1994 and, again, by the post-2008 slow-down in demand. Even greater discontinuities are possible if deliberate management (such as more aggressive efficiency goals for road vehicles) were to shape the profile of future oil demand. If a looming physical shortage of oil were to become a matter of humanity's survival, then clear priorities could ensure an extended period of adequate supply by allocating the refined fuel according to a firm hierarchy of priority uses. Fuel for agricultural machinery, indispensable aviation and feedstock for essential petrochemical syntheses would be in the first category; fuels for long-distance transport of perishable goods in the second (but all railway traffic should be electrified); and gasoline for passenger cars would get the lowest ranking while that transportation sector undergoes gradual conversion to electric drive.

All of this means that we do not know when the global extraction will peak, at what level it will, and if it will be followed by decline mirroring the historic run-up, by a more gradual prolonged retreat or by a substantial drop followed by decades-long fluctuating plateaux. We will not get close to fairly accurate answers until we have explored all of the world's sedimentary basins in great enough detail to offer a narrowly constrained estimate of ultimate reserves. But, as Morris Adelman put it succinctly: 'To know ultimate reserves, we must first have ultimate knowledge. Nobody knows this, and nobody should pretend to know.' But even if we had a perfect knowledge of the world's ultimately recoverable oil resources, the global oil production curve could not be drawn without also knowing the future oil demand.

This is impossible because demand will be driven, as in the past, both by predictable forces (including growing populations and higher disposable income) and by unpredictable political and socioeconomic changes and, above all, by new technical advances. Four prominent historic examples illustrate the repeated extent of our ignorance, with unpredictable outcomes having consequences in either direction, that is, both boosting and depressing future price and future oil supply. In 1930 nobody could have predicted the introduction of commercial jet aircraft by 1960, an innovation that created an entirely new economic sector with a large demand for kerosene. In 1960 nobody could have predicted oil prices rising by an order of magnitude as a result of OPEC's actions, a political shift that, for the first time since the 1860s, led to a notable decline in global oil demand. In the early 1980s, as oil prices set new records, nobody could have predicted that a quarter century later half of the passenger-carrying vehicles in the US would be gasoline-guzzling SUVs, pick-up trucks and vans. And in 2005, as media reveled in reports of the imminent peak of global oil extraction, nobody foresaw that a decade later oil prices would be sliding toward new lows as the world worried about large surpluses of oil supply!

As a result, linear assumptions based on the past rates (be they longer lasting or short-term) are risible. US oil demand rose 50% between 1965 and 1973 but less than 2.5% during the two decades between 1979 and 1999; US oil extraction decreased by 20% between 1995 and 2006, but it increased by about 85%

during the next ten years. Which one of these values should we use for a truly long-range, say at least half a century, forecast of demand and supply? And there is yet another enormous uncertainty: as yet we have no idea to what extent the rising concerns about global warming will affect the future extraction of fossil fuels. Study of energy transitions and realities of inertial, embedded energy uses and infrastructure preclude any rapid abandonment of fossil fuels in general and crude oil in particular.

Even with remarkable technical advances there is no doubt that energy transitions present enormous problems for the providers of energies that are being replaced (OPEC members certainly do not look forward to any early end of the oil era), that they necessitate scrapping or reorganization of many old infrastructures (think of all the oil tankers, pipelines and refineries), and that they require the introduction of entirely new links, procedures and practices (no matter if the dominant new resources are solar or nuclear). Resulting sectoral and regional socioeconomic dislocations are thus inevitable and can be deep and long-lasting (think of the economically depressed former major coal-mining regions), the necessary infrastructural transformations will be costly (valued in trillions of dollars) and inevitably protracted (requiring decades rather than years to put in place) and their diffusion will be uneven (they always have been: even in the US many rural areas were electrified only during the 1950s and about 1.5 billion people worldwide still have no electric lights!).

At the same time, we cannot discount the possibility that concerted global action could accelerate the transition to non-carbon energies, especially as far as electricity generation is concerned — but such a shift would have only a limited impact on the global demand for refined oil products whose principal market is, and will remain, in transportation. A greater impact could result from determined efforts to limit the use of fossil carbon in order to prevent excessive rise of atmospheric temperature caused by the emissions of anthropogenic greenhouse gases. In 2017 about one-third of all CO_2 emissions from commercial energy use originated in combustion of refined oil products (in 1950 it was about one-quarter), a share too large to achieve meaningful reductions of future emissions by concentrating only on other fossil fuels. So far, our efforts have been quite inadequate. The goal of the Paris (COP 21) meeting of November 2015 was to work toward limiting the rise of average tropospheric temperature to no more than 2°C – yet even if all the national pledges submitted at that time were completely fulfilled, the meeting's final document noted, with concern, that the estimated emissions in 2025 and 2030 would not fall within the desired scenarios but rather lead to a further increase of annual CO_2 generation.

Achieving the desired goal would require actual cuts in the current rates of fossil fuel combustion, including substantial reductions of oil use. The chances of ending the fossil fuel era in a matter of two or three decades appear quite unrealistic: in 2017 the world derived about 85% of its primary commercial energy from the combustion of fossil carbon. The coming years will show how far our efforts will go, but it is most likely that we will not stay below 2°C and will have to do our best to adapt to the resulting warming. At the same time, a long-term outlook is more encouraging: in two generations (2060s) we will have made substantial progress toward decarbonizing the global energy supply and although we will still rely on oil as one of the fundamental energizers of our economies we should be doing so with a greatly reduced environmental impact.

Beyond oil

A fundamental general consideration needs to be stressed before I proceed with a brief outline of alternatives to conventional liquid oil. Substitutions that are already technically proven (such as gas-to-liquid conversions) or that appear as highly promising future candidates (for example ethanol production from cellulosic biomass) are nevertheless often seen as unacceptable or impractical simply because they cost (or are projected to cost) more than the conversion they are set to replace. This simplistic cost argument is misleading for three principal reasons. First, it does not acknowledge that the real cost of today's liquid fuels is higher (often substantially so) than the price directly paid by consumers. Second, it implies that only the resources and conversions secured with the lowest cost are worthy of consideration regardless of the environmental or strategic implications of their use. Third, it ignores the fact that modern societies are already paying far less (as a share of disposable income) for their energy needs than at any time in history and hence even a doubling of such vital expenditures would not be catastrophic.

The last point is true even in the country that is most addicted to excessive driving. Detailed surveys of US consumer spending show that in 2015 an average family spent less on gasoline (3.7% of all expenditures) than it did on entertainment (5%) or food away from home (5.4%). Why then should we be panicked by the prospect of an alternative motor fuel that retails (say, arbitrarily) at twice the price of today's gasoline? If that new fuel were to be used in vehicles operating with double the efficiency of today's cars (given the poor average performance of US cars this is an easily achievable goal), then even this low share of expenditure may remain unchanged! And the argument about unacceptably higher costs of alternative fuels is insufficient if they could be produced and converted with lower environmental burdens (including lower emissions of greenhouse gases) or if they will provide important strategic benefits.

Such considerations may drive future efforts to extract even more oil from known reservoirs: after all, even today's best enhanced recovery methods still leave behind 40–50% of the oil originally in place, and higher prices may justify more expensive recovery techniques. Mining of progressively poorer mineral ores is perhaps the best analogy. And beyond the conventional liquid oil there are vast resources of nonconventional hydrocarbons whose recovery is already contributing to most of the crude oil produced in North America (a combination of American shale oil and Canadian oil extracted from oil sands) and will be gradually adopted in other parts of the world rich in such nonconventional oil resources. This is an apposite place to stress that there are no sharp and obviously apparent divides between the two kinds of oil resources and that the same is true for their extraction. In quality terms, this continuum extends from light oils (API gravity of at least 25°) to medium heavy oils (20–25° API, mobile at reservoir conditions) to heavy oils (10–20° API, still somewhat mobile) to bitumen (less than 10° API, non-mobile) and finally to oil shales with minimal or virtually no permeability.

Limited volumes of medium and of some extra heavy oils have been produced for many years in Saskatchewan and Venezuela. Alaska's North Slope also has large deposits of heavy oil (up to 33Gb) and experiments have begun with its commercial production. But most of the world's 4Tb of heavy oils in place are found in Venezuela (close to 1Tb) and in North America's most important commercial concentration, in Alberta's oil sands, with proved reserves of about 166Gb in 2014, making Canada the country with the world's third largest oil reserves behind Venezuela and Saudi Arabia. Total (mining and *in situ*) extraction reached 2.6Mbpd in 2017, which means Alberta's conventional oil contributed less than 15% of the province's total oil extraction. Much like the future of US oil, Canada's oil prospects thus depend heavily on the recovery of a nonconventional resource.

OIL FROM SANDS

Extraction of oil from oil sands was commercialized for the first time on a small scale during the late 1960s. Suncor was the first company to produce oil from Alberta oil sands near Fort McMurray in 1967, and the Syncrude consortium, formed in 1965, has been producing in the area since 1978. Both of these pioneering projects operate large oil sand mines, using huge excavators to mine the rock and the world's largest off-road trucks to transport it to a bitumen extraction plant from which it moves to an upgrading facility to yield light crude oil. Mining of Alberta oil sands, extraction of bitumen and its upgrading to light crude oil returns about six units of energy for every unit invested. Only about a fifth of all recoverable oil in Alberta's oil sands can be reached by surface mining; the rest will have to be extracted *in situ* and two techniques (see figure 28) have been commercialized so far, cyclic steam-stimulation (CSS) and steam-assisted gravity drainage (SAGD).

Imperial Oil's Cold Lake Project was the first CSS recovery that alternates periods of injecting hot pressurized steam (300°C, 11MPa) into well bores with periods of soaking that loosens the bitumen. These cycles last from a few months to three years and the heated bitumen–water mixture is drawn from the same wells that were used for steam injection. On average, this process extracts about a quarter, and with follow-up processes up to 35%, of the bitumen originally present in sands. SAGD was patented by Imperial Oil in 1982 and Cenovus Energy is now its leading practitioner in Alberta oil sands. The process uses two horizontal wells (typically 500–800m long) that are drilled near the bottom of an oil sands formation and are separated by a vertical distance of 4–6m. Steam injected into the top well heats the surrounding bitumen that slowly drains into the bottom well from which the mixture of water and oil is then lifted. Cenovus also injects butane along with steam and adds a horizontal well between a pair of SAGD wells in order to boost recovery to as much as 70%. Separated water is recycled to generate steam, and while the early *in situ* recovery required (in volume terms) nearly eight units of steam to produce a unit of oil, the latest practices have reduced that (and with it the energy cost to produce steam) to two units or even slightly less. Although most companies still operate with EROEI of 3:1, the best recovery practices have doubled that ratio.

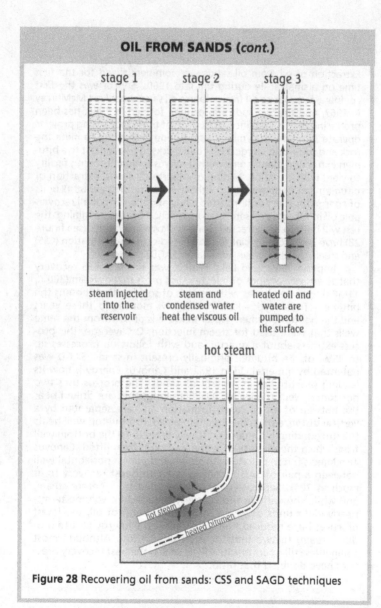

OIL FROM SANDS (cont.)

stage 1

stage 2

stage 3

steam injected into the reservoir

steam and condensed water heat the viscous oil

heated oil and water are pumped to the surface

hot steam

hot steam

heated bitumen

Figure 28 Recovering oil from sands: CSS and SAGD techniques

Petróleos de Venezuela has been exporting heavy boiler fuel (coal or gas substitute), a mixture of 70% natural bitumen, 30% water and a small amount of an additive that stabilizes the emulsion, as liquid *Orimulsión* burned in electricity generating plants, but any large-scale development of the Orinoco Belt's enormous nonconventional oil resources will be determined not only by a complex interplay of oil prices and technical advances but also by the political situation in what has become one of the world's most precarious economies with declining conventional oil extraction (20% down between 2006 and 2015).

Before nonconventional oil became a major addition to the global supply of liquid fuels, there was much interest in gas-to-liquid (GTL) conversions. Their technical feasibility was demonstrated for the first time in 1923 with the production of motor fuels from coal gas by a process invented by Franz Fischer and Hans Tropsch in Germany. During World War II the Fischer-Tropsch synthesis kept the Wehrmacht and Luftwaffe supplied with fuel; the output peaked in 1943 at 124,000bpd before it was reduced by Allied bombing. Inexpensive imported crude oil made further GTL ventures uneconomical, but in 1955 production began at South Africa's Sasol plant. The improved Sasol process was used abroad for the first time in 2007 when Oryx GTL was completed in Qatar. Shell built its first GTL plant in Bintulu in Malaysia in 1993 and in 2012 completed the world's largest project, Pearl GLT in Ras Laffan in Qatar, using natural gas from the world's largest gas field in the Persian Gulf.

But the high cost of these plants and fluctuating, low costs of crude oil have not led to any large-scale adoption of these designs, and in 2017 all operating GTL projects supplied less than half a percent of worldwide liquid fuel production. The ultimate goal of natural gas conversions is to commercialize new methods that would transform methane into affordable energy carriers. George Olah, the Nobel Prize winner in chemistry for 1994, has argued that a methanol would be the best choice. This liquid hydrocarbon (CH_3OH) can be prepared by a number of methods – above all, by direct oxidative conversion of natural gas, and also by catalytic reduction of CO_2 – and it is safer and less expensive to handle than hydrogen, a gas that has been seen by many as the ultimate energy carrier of future non-carbon economies.

But near-term displacement of oil by gas does not have to go through liquids. Direct replacement of liquids by natural gases has been underway for decades and it will continue for decades to come. As already noted, natural gas replaced fuel oil in heating all but a small share of homes both in North America and in Europe, and it has been substituted for fuel oil in many electricity-generating plants and in industrial enterprises. Natural gas is also highly valued petrochemical feedstock. And natural gas can replace motor gasoline in passenger cars in two ways: directly, and with only minor engine modifications, as compressed natural gas suitable for urban fleet operations and other city driving; and indirectly by using highly efficient combined cycle gas turbines to generate electricity for electric cars. Simply put, with the exception of flying and long-distance land and maritime transport (energy density of natural gas under normal pressure is only 1/1000 that of liquid fuel, making it unsuitable as a portable fuel sufficient for extended travel), everything that is done with liquid fuels can be done with gases.

NATURAL GASES: PROPERTIES AND RESERVES

Much like oils, natural gases are mixtures of variable proportions of hydrocarbons, but unlike oils they are primarily mixtures of just three of the simplest alkanes: methane, ethane and propane. Higher homologues (butane, pentane and hexane, are separated as natural gas liquids) and CO_2, H_2S, N, He, and water vapor, found in many gases, are also separated before the gases are compressed and transported by pipelines. Natural gases are commonly associated with crude oils but they also exist as free (dry) gases without any contact with crude oil in an oil reservoir or in entirely separate

NATURAL GASES: PROPERTIES AND RESERVES (cont.)

gas-bearing formations. Their heat content ranges between 30 and 45MJ/m³ (35.5MJ/m³ for CH_4), they are the least polluting fossil fuels, and they generate the least amount of CO_2 per unit of energy. As with crude oil, conventional gas reserves have been steadily increasing and by 2017 they had reached 190Tm³, or in energy terms nearly as much as the total 2016 reserves of conventional crude oil.

This increase has not only accommodated the expanding extraction (it rose more than 2.1-fold between 1985 and 2016, to 3.5Tm³) but it maintained the global R/P ratio at more than fifty years. Conventional reserves are concentrated in Iran (about 18% of the total), Russia (about 17%), Qatar (about 13%) and Turkmenistan (nearly 10% of the total). The Middle East claims just over 40% of the global total, much less than its share of crude oil. Gas associated with oil used to be simply flared as an unwanted by-product but this wasteful practice has declined with the rising demand for clean household and industrial fuel. In 1975, gas equal to about 14% of worldwide production was flared, with major sites visible on nighttime satellite images as lights brighter than those of many large cities. Worldwide flaring has been slowly declining since the beginning of the new century but in 2015 it had a slight uptick and it still accounts for slightly more than 4% of global production (more than China's annual output in that year!), with Russia, Iraq, Iran, the US (due to rapidly expanding shale oil extraction) and Venezuela being the top offenders.

For decades those large natural gas reserves that could not be accessed by a pipeline could not be used, creating large stores of so-called stranded gas. This limitation began to change only during the early 1960s when the first liquefied natural gas (LNG) tankers were used to export Algerian gas to the UK and France, and Indonesian gas to Japan. But the gas liquefaction plants and special tankers were expensive and the export remained limited. LNG occupies only about 1/600 the volume of natural gas but it must be cooled to −162°C and regasified after delivery. As both the liquefaction process and the construction of special tankers with highly insulated containers became more affordable, LNG trade took off during the 1990s and its unfolding expansion has finally changed the natural gas trade into a truly global endeavor with substantially lower prices and a growing number of suppliers and buyers. In 2017 eighteen countries were exporting LNG from more than 70 liquefaction plants (with Qatar, Malaysia, Australia and Nigeria in the lead) and thirty-nine countries were importing the fuel, with all of the largest buyers (Japan, South Korea, China and India) in Asia. Nearly 400 LNG tankers were used in the exports, the largest ones with capacities of more than 250,000m^3, and 105 of them carried almost 33% of all traded gas.

Prospects for long-term supply of natural gas look highly promising. In 2000 the US Geological Survey put cumulative production of conventional natural gas at 292 billion barrels of oil equivalent (Gboe) and remaining reserves at 800Gboe, and assumed reserve growth of about 610Gboe and undiscovered reserves at 866Gboe for the total EUR of 2.57Ttoe, only about 15% less than EUR of crude oil. All of these figures refer to conventional resources only. The only nonconventional gas resource that has been exploited for decades is coalbed methane, but since 2005 the US production of shale gas, which resulted in a 50% increase of output by 2015, showed that hydraulic fracturing can transform the industry as much as it has transformed he extraction of crude oil. The US Energy Information Administration put the undiscovered but technically recoverable resources of shale gas at more than 210Tm^3, slightly larger than the existing reserves of conventional gas. The countries with the largest shale gas potential are China, Argentina, Algeria, the US, Canada and Mexico.

As large as shale gas resources may be, they are insignificant when compared to methane hydrates (clathrates) that were formed by the gas released from anoxic decomposition of organic sediments by methanogenic bacteria and are now trapped inside rigid lattice cages of frozen water molecules. Fully saturated gas hydrates have one CH_4 molecule for every 5.75 water molecules and hence $1m^3$ of hydrates contains as much as $164m^3$ of CH_4. Two environments favor the formation of hydrates: polar continental sediments (at depths between 100m and 2.5km) and sediments beneath the ocean floor in many latitudes. More than 200 gas hydrate deposits have been identified worldwide, including those in the Russian, American and Canadian Arctic, off California and Guatemala and in Japan's deep (water depths about 4,700m) Pacific Nankai Trough (hydrates about 4,800m below the sea bottom). The resource base of methane hydrates is so immense that only gross approximations of its size are now possible, and even very conservative estimates indicate the enormity of resources in place. The global total may be as much as 10Tt, or about twice as much as all carbon stored in coals and conventional hydrocarbons, while hydrates below the seabed of American coastal waters may be three orders of magnitude more voluminous than the country's conventional gas reserves.

The first small-scale production (decompression and heating) trials were done in Canada in 2001 and 2002, but abundant shale gas supplies ended further experiments. In 2013 the Japanese experimented with the recovery of Nankai Trough hydrates and obtained good flow rate before the pumps became clogged by incoming sand, and in 2017 the Chinese conducted successful recovery tests in the South China Sea. No commercial breakthrough will come anytime soon and any large-scale developments will have to take into account the possibility of a sudden catastrophic release of some hydrate deposits into the atmosphere, a most unwelcome contribution to anthropogenic greenhouse gas emissions. But these challenges are hardly a valid argument for eliminating this enormous resource from future consideration as a major source of hydrocarbons (perhaps even by the middle of this century): doing so would be akin to claiming in 1930 that any extraction of oil from offshore fields out of sight of land was impossible, or to concluding in 1950 that no oil or gas would ever be produced from shales.

Greater use of natural gas can lower our reliance on crude oil in many economic sectors. Abundant and affordable natural gas can eliminate all remaining uses of crude oil and refined oil products in electricity generation and space heating; vehicles used in city traffic could easily be converted to run on compressed natural gas; trucks and LNG trucks could be powered by liquefied gas; and many industrial enterprises relying on refined products for process heat or hot water can generate those needs with higher efficiency by switching to gas. Highly efficient combined cycle gas turbines could produce electricity to power electric cars, and their generation costs could be lowered by integrating them with renewable conversions in sunny and windy places. At the same time, it will be a much greater challenge to reduce further, and eventually eliminate, our dependence on refined fuels in shipping and flying.

That is why any long-term assessment of oil's futures must consider both the supply prospects and the imperatives of demand. As for the supply, we have still formidably large (and still growing) conventional crude oil reserves, at least as large (but most likely considerably larger) technically recoverable reserves of oil and natural gas in shales and improving capabilities of gas-to-liquid conversions: this combination means (even when assuming no early breakthroughs in hydrate recovery) that hydrocarbons should remain the leading source of global commercial energy supply far beyond the middle of this century. On the demand side, many existing uses of refined oil products could be replaced by natural gas or by primary (renewably generated) electricity, but it is most unlikely that by 2050 we will see the entire global car and truck fleet operating without gasoline and diesel, and the prospects for concurrently eliminating diesel fuel and kerosene from, respectively, shipping and air transport are considerably less.

In a speech accepting the Biennial OPEC Award for 2006, the late Peter Odell, one of the most astute, life-long, observers of the global oil scene, concluded that 'peak-oilers', much like their numerous predecessors, will soon be proven wrong, that the present contribution of oil and gas to the global energy supply will be only modestly reduced by 2050 and that natural gas will surpass oil as the leading source of primary energy. At that time, I concurred with Peter in my writings, and more than a decade later I still see such conclusions as the most sensible qualitative forecasts of the world's energy futures (as always, I stay away from any specific quantitative details, as such long-term predictions are beyond our capability).

And as the importance of this mixture of liquid and gaseous and conventional and nonconventional hydrocarbons recedes, first in relative and later also in absolute terms, renewable energy flows will be gaining higher shares of global primary energy supply. Liquids from biomass (ethanol from sugar cane, biodiesel from oil seeds) are already displacing some gasoline and a small share of diesel fuel. But the future should not belong to the currently heavily promoted (but in many ways problematic) conversion of grains (above all corn) to ethanol and oil seeds to biodiesel, but to innovative bioengineering processes converting more abundant cellulosic biomass, including the share of crop residues (mainly straw) that does not have to be recycled to maintain soil quality and high-yielding perennial grasses planted on non-agricultural land.

Wind-powered electricity generation, onshore and offshore, still accounts for only a small part of global generation (about 4% in 2016) but it has already made major strides in a number of European countries and larger turbine sizes (in 2016 up to 8MW) and higher capacity factors (some offshore wind farms can generate electricity for up to 47% of the time) will raise its contributions for many years to come. In 2016, photovoltaic conversion of solar radiation contributed even less to the global electricity generation total (just over 1%) but its numerous advantages (quiet, no moving parts, durable, high power density), falling costs and improving efficiencies guarantee its continued future expansion. Although these two conversions may seem to do little for expanding the supply of liquid fuels they could actually fit perfectly into a system of rechargeable cars, be they hybrids or pure electrics.

The inherent fluctuations and unpredictability of photovoltaic electricity generation in temperate latitudes would be much less of a problem for recharging plug-in hybrid cars than for lighting houses or running machines in a factory where electricity must be available on demand. In contrast, a car, be it in a garage or in a parking lot at a place of work, could be recharged whenever a surge of renewable electricity became available.

Regardless of the actual rate of oil extraction and the eventual date of the highest annual production of oil from any resources, there is no reason to see the transition to the post-oil era as a period of unmanageable difficulties or outright economic and social catastrophes. Historical evidence is clear: energy transitions have always been among the most important stimuli of technical advances (think of new prime movers, new materials and new energy converters), promoting innovation (such as the profound managerial and organizational changes brought by computers), higher efficiency (for example, a gas turbine vs. steam engine) and resource substitution (like the substitution of coke made from coal for charcoal from disappearing forests during the late eighteenth and nineteenth centuries). Their outcomes – coal replacing wood, oil replacing a great deal of coal, now natural gas already replacing a great deal of oil – have shaped modern industrial, and post-industrial, civilization, leaving deep imprints on the structure and productivity of economies as well as on the organization and the quality of life of affected societies.

There is no doubt that the unfolding energy transition (whose eventual outcome would be the replacement of fossil fuels by non-carbon energies) is extraordinarily challenging, mainly due to the scale of our reliance on fossil fuels, cost and inertia of enormous infrastructures required for its reliable functioning, and the continuing concentration of humanity not just in cities but in megacities containing more than 10 million people. As with all long-range perspectives, it is counterproductive to pinpoint any rates of progress, dates or shares for new energy conversions that will be needed to accomplish the epochal shift. This we know for certain: although a small nation can switch from one dominant form of energy to another in a matter of years, energy transitions are normally protracted affairs, extending across decades rather than years. And neither the tempo nor the eventual achievements of these long transitions from first commercial uses to widespread embrace to eventual domination can be judged by the state of affairs during the initial stages of expansion. Precisely because this transition will have to be gradual and protracted, technical means available two or three generations from now may provide effective solutions for many of today's intractable problems.

Energy transitions – from biomass to coal, from coal to oil, from oil to natural gas, from direct use of fuels to electricity – have stimulated technical advances and driven our inventiveness. Inevitably, they bring enormous challenges for both producers and consumers, necessitate the scrapping or reorganization of extensive infrastructures, are costly and protracted and cause major socioeconomic dislocations. But they have created more productive and richer economies, and modern societies will not collapse just because we face yet another of these grand transformations. The world beyond oil is still several generations away but we should see the path toward that era as one of very challenging, but also immensely rewarding, opportunities as modern civilization eventually severs its dependence on fossil carbon.

Appendix A: Units, abbreviations and their definitions and conversions

b	barrel	unit of volume	42 gallons; 158.98l
bpd	barrels per day		
c	centi	multiplier prefix	10^{-2}
dwt	deadweight ton	unit of capacity	1,016kg
ft	foot	unit of length	0.3m
m^3	cubic meter	unit of volume	1,000l; 264.17 gal
g	gram	unit of mass	
G	giga (billion)	multiplier prefix	10^9 (thousand million)
Gal	gallon (US)	unit of volume	3.785l
Gb	giga (billion) barrels		
Gboe	giga (billion) barrels of oil equivalent		
Gt	giga (billion) tonnes		
hp	horsepower	traditional unit of power	(1hp = 745.7W)

in	inch	unit of length	2.5cm
J	joule	unit of energy	
k	kilo	thousand, multiplier prefix	10^3
kg	kilogram	unit of mass	1,000g
km	kilometer	unit of length	1,000m
L	liter	unit of volume	$1,000ml^3$; $0.001m^3$
m	meter	unit of length	100cm; 3.28ft
m	milli	multiplier prefix	10^{-3}
M	mega	million, multiplier prefix	10^6
Mbpd	million barrels per day		
MPa	million pascals		
mpg	miles per gallon		
Mt	million tonnes		
Pa	pascal	unit of pressure	
t	tonne (metric ton)	unit of mass	1,000kg
T	tera	multiplier prefix	10^{12} (million millions)
Tt	trillion tonnes		
W	watt	unit of power	
Ωm	Ohm-meter	unit of resistivity	

Appendix B: What's in a barrel? Basic oil properties and measures

Differences in oil composition result in specific densities that range mostly between 0.8 and 0.9 grams per milliliter (g/ml, or 800–900kg/m³) but whose extremes are about 0.74 and 1.04g/ml, or as light as a gasoline and slightly heavier than water. This means that one barrel, the standard non-metric volume measure commonly used in the oil industry, has no single mass equivalent. This nineteenth-century container with the volume of 42 US gallons (or roughly 159 liters) was adopted by the US Bureau of the Census in 1872 and it has since been used worldwide for both production statistics and in evaluating oil resources; its common abbreviation is either bbl (for blue barrel) or b (I use the latter in this book). With heavy crude oils, just over six barrels weigh a metric ton (t), whereas with light oils up to 8.5 barrels are needed; most oils fall between 7 and 7.5b/t, and the value of 7.33b/t is the most commonly used average. Because international oil production and reserves continue to be reported in barrels I will use scientific prefixes for their multiples (M for millions, G for billions and T for trillions). Production in barrels per day will be abbreviated as bpd.

Barrel is not the only non-metric unit used by the oil industry. Crude oil densities are designated in terms of degrees of the American Petroleum Institute (API) gravity based on an arbitrary assignment of 10° API to water; conversion from specific density thus follows the general formula $°API = (141.5/\text{specific density})-131.5$. Consequently, a very heavy crude oil with specific density of 0.95 (close to that of water) has $°API$ 17.5, while a light oil with specific density of 0.82 has $°API$ 41. Crude oils with densities above $°API$ 31.1 are classified as light; heavy oils have $°API$ below 22.3. All liquid crude oils are lighter than water and are also immiscible with it and float on its surface: only a strong agitation will produce oil-water emulsion.

Crude oils that dominate the global trade are mostly of medium density or are only moderately light. Most Saudi crude oils rate between $°API$ 28–33, oil from Kuwait's largest oilfield has $°API$ just over 23, and the southern Iraqi oil from Basra is rated about $°API$ 25. Crude oil from Alaska's North Slope has $°API$ 29 but the North Sea Brent oil has about $°API$ 38. Some Libyan, Algerian and Nigerian oils are very light, with $°API$ 37–44, and the lightest internationally traded crude oil, with $°API$ 60, comes from Australia's North-Western Shelf. Price is the most obvious marker of crude oil densities: generally, the lighter the oil (the higher the $°API$), the higher its price.

Appendix C: Short glossary

alluvial fan	poorly sorted rock debris deposited by rivers on flat land
anthracite	highest-quality coal, nearly pure carbon
anticline	convex (arch-like) fold of stratified rocks
biomass fuel	any fuel derived from plants: wood, charcoal, crop residues (mostly straws), also dried dung and biogas
bottomhole assembly	the lower portion of the drill string made up (from the bottom up) of the drilling bit, drill collar, heavy drill pipe and crossover pipes
Carboniferous period	359–299 million years ago, known for its coal deposits
catagenesis	a process that converts organic kerogens in hydrocarbons
Cenozoic era	from 66 million years ago to the present
compression ratio	ratio of the maximum and minimum volume of gas in an engine's cylinder; compares the volume of gas when the piston is at the top of the stroke to the volume of gas when the piston is at the bottom of its stroke
Cretaceous period	145–66 million years ago, part of the Mesozoic era known for giant dinosaurs
derrick	a pyramidal structure to support a set of pulleys (crown block) and the drill string of a drilling rig

diapir	a geological structure formed by the upward flow of material into brittle surrounding rocks
drill pipe	hollow, thin-walled steel pipe used in oil and gas exploration and production
estimated ultimate recovery	total amount of oil or gas that can be recovered from a well or a field
fracking	see hydraulic fracturing
hydraulic fracturing	pumping of pressurized fluid (mostly water and sand) into a well bore to crack (fracture) surrounding rocks and stimulate oil or gas production
hydrocarbons	collective term for crude oils and natural gases
hydrogen-to-carbon ratio	atomic ratio of the two elements in a compound; a higher ratio means more energy release on combustion
induction log	measurement of formation resistivities in a borehole
Jurassic period	201–145 million years ago, part of the Mesozoic era
kelly	a long steel bar (hexagonal or square, with a hole in the center) used to transmit rotation from the rotary table to the drill string
kerogen	heavy organic matter in sedimentary rocks that is insoluble in organic solvents
lignite	poorest-quality coal containing lots of water and ash
Mesozoic era	252–66 million years ago
metamorphic rock	created by transformation (by heat and pressure) of a pre-existing rock

Paleocene epoch	66–56 million years ago, the oldest part of the Cenozoic era
Palaeozoic era	541–252 million years ago
phytomass	plant mass (trees, crops, grasses, aquatic plants)
recovery factor	recoverable amount of oil as a proportion of oil originally in place
resistivity	ability of a material to resist electrical conduction
reserve/production ratio	remaining amount of a non-renewable resource, current rate of extraction
prime mover	natural or mechanical source of kinetic energy
sandstone	sedimentary rock composed mostly of quartz
shale	sedimentary rock composed of clay and other minerals
shelf	edge of a continent under shallow ocean
stratigraphy	geological discipline studying the order, scale and age of rock layers
substrates	soils, sediments, mineral and bedrock formations
trap	a configuration of rocks that contains oil and gas sealed by impermeable layers
tripping	pulling the drill string from a well bore and returning it
well log	measurement of specific physical variables versus depth or time during drilling

For a detailed glossary of oil drilling and oilfield terms see:

Schlumberger. Oilfield glossary www.glossary.oilfield.slb.com

Appendix D: Additional reading and websites

Books

General

Bridge, G. and P. Le Billon. 2013. *Oil*. Cambridge: Polity Press.
Maugeri, L. 2006. *The Age of Oil: The Mythology, History, and Future of the World's Most Controversial Resource*. Westport, CT: Praeger Publishers.
Yergin, D. 2012. *The Quest: Energy, Security, and the Remaking of the Modern World*. New York: Penguin.

Geology and resources

Ahlbrandt, T. S. et al. 2005. *Global Resource Estimates from Total Petroleum Systems*. Tulsa, OK: American Association of Petroleum Geologists.
Li, G. 2011. *World Atlas of Oil and Gas Basins*. Chichester: Wiley-Blackwell.
Selley, R. C. 1997. *Elements of Petroleum Geology*. San Diego, CA: Academic Press.

History of oil industry

Brantly, J. E. 1971. *History of Oil Well Drilling*. Houston, TX: Gulf Publishing.
Howard, R. 2008. *The Oil Hunters: Exploration and Espionage in the Middle East*. London: Hambledon Continuum.
Yergin, D. 2008. *The Prize: The Epic Quest for Oil, Money and Power*. New York: Free Press.

Oil market and prices

Aguilera, R. F. and M. Radetzki. 2015. *The Price of Oil*. Cambridge: Cambridge University Press.

Cordesman, A. H. and K. R. al-Rodhan. 2006. *The Global Oil Market: Risks and Uncertainties*. Washington, DC: CSIS Press.

Marcel, V. 2006. *Oil Titans: National Oil Companies in the Middle East*. London: Chatham House.

OPEC

Ghanem, S. M. 2016. *OPEC: The Rise and Fall of an Exclusive Club*. London: Routledge.

OPEC. 2017. *Annual Statistical Bulletin*. Vienna: OPEC.

Ramady, M. and W. Mahdi. 2017. *OPEC in a Shale Oil World: Where to Next?* Berlin: Springer Verlag.

Peak oil

Campbell, C. J. 2005. *Oil Crisis*. Brentwood: Multi-Science Publishing.

Deffeyes, K. S. 2001. *Hubbert's Peak: The Impending World Oil Shortage*. Princeton, NJ: Princeton University Press.

Lynch, M. C. 2016. *The 'Peak Oil' Scare and the Coming Oil Flood*. Santa Barbara, CA: Praeger.

Hydraulic fracturing

Drogos, D. L., ed. 2016. *Hydraulic Fracturing: Environmental Issues*. Oxford: Oxford University Press.

Gold, R. 2014. *The Boom: How Fracking Ignited the American Energy Revolution and Changed the World*. New York: Simon & Schuster.

Smith, M. B. and C. Montgomery. 2015. *Hydraulic Fracturing*. Boca Raton, FL: CRC Press.

Future of oil

Odell, P. R. 2004. *Why Carbon Fuels Will Dominate the 21st Century's Global Energy Economy.* Brentwood: Multi-Science Publishing.

Olah, G. A., Goeppert, A. and G. K. S. Prakash. 2006. *Beyond Oil and Gas: The Methanol Economy.* Weinheim: Wiley-VCH.

Smil, V. 2017. *Energy Transitions: National and Global Perspectives.* Santa Barbara, CA: Praeger.

Websites

American Association of Petroleum Geologists: www.aapg.org

American Petroleum Institute: www.api.org

British Petroleum: www.bp.com

British Petroleum Statistical Review of World Energy: www.bp.com

Canadian Association of Petroleum Producers: www.capp.ca

Chevron Corporation: www.chevron.com

Exxon Mobil: www.exxonmobil.com

OPEC: www.opec.org

Platts Oil: www.platts.com

Rosneft: www.rosneft.com

Royal Dutch Shell: www.shell.com

Saudi Aramco: www.saudiaramco.com

Schlumberger: www.slb.com

Schlumberger Oilfield Glossary: www.glossary.oilfield.slb.com

US Energy Information Administration, international statistics: www.eia.doe.gov

World Oil: www.worldoil.com

Index

References to images are in *italics*.